NG-84

1990

Recreational Uses of Coastal Areas

The GeoJournal Library

Volume 12

Series Editor: Wolf Tietze, Helmstedt, F.R.G.

Editorial Board: Paul Claval, France
R. G. Crane, U.S.A.
Yehuda Gradus, Israel
Risto Laulajainen, Sweden
Gerd Lüttig, F.R.G.
Walther Manshard, F.R.G.
Osamu Nishikawa, Japan
Peter Tyson, South Africa

The titles published in this series are listed at the end of this volume.

Recreational Uses of Coastal Areas

A Research Project of the Commission on the Coastal Environment, International Geographical Union

Edited by

PAOLO FABBRI

*Department of Historical Disciplines,
University of Bologna, Italy*

Kluwer Academic Publishers
Dordrecht / Boston / London

Library of Congress Cataloging-in-Publication Data

```
Recreational uses of coastal areas : a research project of the
  Commission on the Coastal Environment, International Geographical
  Union / edited by Paolo Fabbri.
      p.   cm. -- (The GeoJournal library)
    Bibliography: p.
    ISBN 0-7923-0279-6 (U.S.)
    1. Coasts--Recreational use.  I. Fabbri, Paolo, 1948-
  II. International Geographical Union. Commission on the Coastal
  Environment. III. Series.
  GV191.66.R4 1989
  333.78'4--dc20                                              89-8016
```

ISBN 0-7923-0279-6

Published by Kluwer Academic Publishers,
P.O. Box 17, 3300 AA Dordrecht, The Netherlands.

Kluwer Academic Publishers incorporates
the publishing programmes of
D. Reidel, Martinus Nijhoff, Dr W. Junk and MTP Press.

Sold and distributed in the U.S.A. and Canada
by Kluwer Academic Publishers,
101 Philip Drive, Norwell, MA 02061, U.S.A.

In all other countries, sold and distributed
by Kluwer Academic Publishers Group,
P.O. Box 322, 3300 AH Dordrecht, The Netherlands.

printed on acid free paper

All Rights Reserved
© 1990 by Kluwer Academic Publishers
No part of the material protected by this copyright notice may be reproduced or utilized in any form or by any means, electronic or mechanical, including photocopying, recording, or by any information storage and retrieval system, without written permission from the copyright owner.

Printed in the Netherlands

Table of contents

Introduction
Paolo Fabbri vii
Acknowledgements xix

SECTION I: REGIONAL STUDIES

1. The recreational use and abuse of the coastline of Florida
   R. W. G. Carter 3
2. Management strategies for coastal conservation in South Wales, U.K.
   Allan T. Williams 19
3. Recreational uses and problems of Port Phillip Bay, Australia
   Eric Bird and Peter Cullen 39
4. Recreation in the coastal areas of Singapore
   P. P. Wong 53
5. The Azov Sea coast as a recreational area
   V. A. Mamykina and Yuri P. Khrustalev 63
6. The influence of ethnicity on recreational uses of coastal areas in Guyana
   V. Chris Lakhan 69
7. Recreational uses in the coastal zone of central Chile
   Consuelo Castro and M. Inés Valenzuela 83
8. Recreational uses of Québec coastlines
   Jean-Marie M. Dubois and Marc Chênevert 89

SECTION II: COASTAL RECREATION IN ADVERSE ENVIRONMENTS

9. Recreational use of the Washington State coast
   Maurice L. Schwartz 103
10. Pacific coast recreational patterns and activities in Canada
    Philip Dearden 111
11. The recreational use of the Norwegian coast
    Tormod Klemsdal 125
12. Patterns and impacts of coastal recreation along the Gulf coast of Mexico
    Klaus J. Meyer-Arendt 133
13. Wetlands recreation: Louisiana style
    Donald W. Davis 149

14. The natural features of the Caspian Sea western coasts in the context of their prospective recreational use
   O. K. Leontjev, S. A. Lukyanova and L. G. Nikiforov   165

SECTION III: PLANNING FOR RECREATION

15. Construction of a recreational beach using the original coastal morphology, Koege Bay, Denmark
   Niels Nielsen   177
16. Tourist planning along the coast of Aquitaine, France
   Paolo Ghelardoni   191
17. Sydney's southern surfing beaches: characteristics and hazards
   A. D. Short and C. L. Hogan   199
18. Twenty five years of development along the Israeli Mediterranen coast: goals and achievements
   Yaacov Nir and Avi Elimelech   211
19. Differential response of six beaches at Point Pelee (Ontario) to variable levels of recreational use
   Placido D. La Valle   219
20. Anthropogenic effects on recreational beaches
   Yurii V. Artukhin   231
21. Formulating policies using visitor perceptions of Biscayne National Park and seashore
   Stephen V. Cofer-Shabica, Robert E. Snow and Francis P. Noe   235

SECTION IV: MISCELLANEOUS

22. Marine recreation in North America
   Niels West   257
23. Beach resort morphology in England and Australia: a review and extension
   Dennis N. Jeans   277

Introduction

Human clustering in coastal areas

The coastal zone has gained a solid reputation as a place vocated for recreational activities and this is generally related to the presence of the sea. The relationship, however, does not appear univocal or simple: the sea can be perceived as a hostile element by humans and the more general question of whether the presence of the shore is in itself a favourable, repulsive, or irrelevant factor to settlement is a debatable point, at least for pre-industrial societies.

Back in the early part of the 19th century, Friedrich Hegel regarded oceans and rivers as unifying elements rather than dividing ones, thus implying a trend towards the concentration of human settlements along them. 'The sea', he wrote, 'stimulates courage and conquest, as well as profit and plunder',[1] although he realized that this did not equally apply to all maritime peoples. In Hegel's view, different approaches to the sea were mainly the results of cultural factors and, in fact, he recognized that some people living in coastal areas perceive the sea as a dangerous and alien place and the shore as a *finis terrae*.

However, the precise role of culture in developing such differences still remains unstated and the fact that habitats favoured by living primates other than man are rarely along the shore, suggests that only recently, in the time span of hominid evolution, oceans have been perceived as media for human dispersal. This may also apply to the coastal environment, for which there are signs that, within the same cultural group, approaches might vary in relation to changes in needs. For example, the carbon dating of shell middens, commonly found along many shorelines of the world, shows a more recent age than hunting cultures, suggesting 'that man turned to coastal sources for food only after sources of terrestrial ecosystems were no longer adequate'.[2]

In recent times, a French geographer familiar with many coastal regions of the world, Pierre Gourou, has regarded physical factors as irrelevant to coastal settlement:[3] he mentions 'good coasts' (from a physical point of view) which support dense settlement, with marine activities (Brittany) or without them (Gulf of Guinea); and 'good coasts' which are scarcely populated, as in Southern Chile, Korea, and Kampoutchea. According to Gourou, there are also 'bad coasts' which are heavily settled, with or without economic links to the sea: examples of these are Lebanon and Israel in the first case, and most of China in the latter. 'This field of study', says Gourou, 'is a muddle of contradictions and the only possible explanation lies in culture....' If these are sea-oriented, coastal clustering is likely to develop and nearshore areas will be subject to various degrees of anthropization, conforming to cultural and economic needs and according to technology.

This leaves the question open as to how and why human cultures become sea-oriented or cease to be so. Gourou claims that being oriented seaward or landward depends, for a culture, 'on circumstances which, from time to time, take into account coastal conditions, without being conditioned from them'. Such circumstances, however, are not explicit and this approach indeed seems an excess of reaction to physical determinism. High-latitude coastal areas are practically uninhabited, *mainly* (if not only) as a result of physical factors and this also refers in general to coasts bordering arid regions at all latitudes and to many others characterized by such features as marshlands or a rugged morphology. In all of these circumstances, which are certainly not unusual on this planet, physical features do play a major role in determining densities and patterns of human settlement.

It should be recognized, however, that such physical factors as the presence of freshwater, good (but not excessive) drainage, and soil fertility, which have led the agricultural revolution in its spatial diffusion, as well as the resulting settlement, show no connection whatsoever with the vicinity of the sea. Although specific coastal features may influence the typology and formal distribution of agricultural settlement (for example, linear alignments along the shore, or at a certain distance from it), they are irrelevant as to the density of it, when compared to inland areas. Also, if higher concentrations of settlement and population are to be related – both in space and time – to better opportunities of resource exploitation, such activities as mining, stock raising, or hunting (which imply thin permanent settlement, or none at all) also show no relation to coastal areas, in their spatial diffusion. In a more general way, production of goods and services also does not imply spatial correlations *a priori*, in its distributional patterns, with coastal areas: large urban and industrial areas have been located and developed regardless of distances from the sea. In considering different forms of resource exploitation and the resulting economies, fishing appears as the only long-pursued activity with an obvious link to the coast, or, better, to some kind of surface waters.

On account of its long-recorded history, the Mediterranean region can be used as a good example to review coastal vs. inland trends of settlement and to show how these are led both by environmental features and by a blend of cultural factors. The trend of concentrating settlement and urbanization along coasts in ancient times, is thoroughly documented for the whole basin, and this was primarily due to physical factors. In North Africa and along the eastern side – although inland areas were not as arid as in present times – coastlands certainly enjoyed better conditions for the practice of agriculture. Similar advantages pertained to the coasts on the European side, scattered by marshlands or fringed by steep and rocky cliffs, but also including long stretches of well drained alluvial plains, which provided very favourable habitats for a number of crops. Some of these, like olive and vine cultivation, were eventually extended along sunny slopes to meet ideal growing conditions. To the contrary, inland areas generally presented less favourable conditions, with rugged mountains and extended calcareous formations, resistant to tillage and scarce of surface waters. These areas were thus left for grazing, while agriculture and consequent settlement tended to cluster along the coast.

The other main reason for coastal concentration in the Mediterranean derived from the sea itself: a group of relatively small and contiguous water basins, which were immune from major storms and high tidal ranges and which were crowded with natural harbour sites and landmarks to aid navigation. Even for low-technology cultures, the Mediterranean is an easy sea and, since very early times, it was perceived and used as an extension of the inhabited world, both for fishing and trading. These activities well complemented the farming and grazing economy of riparian settlers and trading was the main stimulus for the expansion and establishment of new colonies by such sea-oriented peoples as the Phoenicians, the Punics, and the Greeks. Which factors pushed these peoples more than others towards the sea? The spirit of adventure and a thirst of profits seem much vaguer than other, more tangible factors such as overpopulation in their narrow coastal or insular homelands; or a more developed and better rooted confidence in the sea, deriving from such natural assets as coastal indentations and the proximity of numerous islands, encouraging short-range navigation.

The political unity of the area, which was attained at the peak of the Roman empire and further maintained in the eastern part of the basin under Byzantine rule, was also a factor favouring flows of goods and peoples across the Mediterranean, thus encouraging coastal settlement. Imperial navies were efficient enough to control all main trading routes and prevent piracy. In fact, the decline of the Roman rule, which can be traced back to the late 3rd century, also marked the first major and recorded decline in the use of the sea. Along with the empire and its economic framework, a network of trading routes and ports was discarded and the sea gradually became less safe for navigation, due to the lack of control and a consequent upsurge of piracy.

In this new situation, coastal settlement became demotivated and dangerous and a landward migration trend was started: this was also encouraged by the upheaval brought to many littoral plains by floods, which, in turn, were caused by deficiencies in water control, and deforestation along slopes used as new farming or grazing areas. Also, hydraulic deregulation created ideal habitats for the endemy of malaria in the resulting marshes. Thus, environmental factors along with political and economic circumstances determined the first withdrawal of the Mediterranean peoples from coasts to inland areas, dispersed coastal settlement, and diverted landward cultural attitudes and resource exploitation.

This process was not fully registered in the eastern part of the region, where no sharp discontinuity is recorded between the Romans and the Byzantines in maintaining a unified political and economic framework. In this area and as far west as the Adriatic, a coastal urban network persisted up to the time of the Arab conquest (7th–8th centuries AD). For the next thousand years, however, the Mediterranean, deprived of any political unity, became essentially an area of conflict between two large cultural groups, the Christians and the Muslims, for which differences in religious beliefs and practices also extended to – and resulted from – different forms of resource exploitation. Over this long period, we can roughly spot two moments of Muslim supremacy (7th–11th and 15th–17th centuries) and one of Christian domination in between, but the balance of power certainly shows more numerous shifts when considered in more detail.

In all cases, the trend of higher/lower coastal settlement, referred to inland areas, has shown a clear and constant dependency on the situation of expansion vs. withdrawal on the adjacent sea. The history of the Mediterranean shows that a push of a people (not necessarily grouped into an organized nationality) towards the sea constitutes a basic premise for that people to restore old ports or build new ones, expand or establish coastal cities, develop a coastal network of communications (on land and/or on sea), reclaim lowlands, and regulate coastal waters in general. This push may be stimulated by overpopulation, leading to a search for new resources on the sea or over it; or by the acquisition of a higher warfare technology; or finally it may be motivated by religion. But the reverse is not necessarily likely to happen: in other words, such facts as overpopulation, technological progress, or religious and ideological tensions do not necessarily lead to a seaward and across-sea-expansion. In any case, factors that lead to a push do not always play the same role. When, between the late 1600s and the early 1800s, the balance of power in the Mediterranean shifted in favour of Western Europe, the religious factor was irrelevant, the demographic one was important, technologically the decisive.

Although not as richly documented as in the Mediterranean, trends to coastal concentration or dispersal have been individuated in many other regions of the globe, always showing close connections to spatio-temporal approaches of human groups towards the sea. There are indeed many situations of densely populated coastlands, as in the monsoon area, where a combination of excellent farming conditions in drained and fertile coastal plains, difficult access to the sea, and low technologies have encouraged a land-based coastal settlement, with densities as high as 700–800 inhabitants/sq. km. Things are rapidly changing, however, in this area too, as technology and overpopulation exert pressures for more sea-oriented cultural approaches.

In the industrial age, two basic facts may be recognized and documented in their entities and distribution:
– a marked and globally diffuse higher concentration of residents – and thus settlements – in the vicinities of the coast;
– a close spatio-temporal relationship between this trend and the spreading of the Industrial Revolution.

A few figures may illustrate the first fact. In the United States, nearly 50% of the total population resides in coastal counties and about 90% of the population growth in the 1970s has taken place in the 30 states facing the oceans or the Great Lakes. In Britain, the coastal population density is 150–180, in contrast with 40–80 inland. In France, littoral townships, with a total area of 3.8%, contain 12% of the total population. In the Iberian Peninsula, 60% of the population lives in littoral provinces.

As to the relationship between coastal concentrations and the Industrial Revolution, this may be evidenced *in fieri* but will be briefly discussed here on an *a posteriori* basis. Which are the processes and facts that link coastal clustering with the development of industrial culture? It should be firstly recalled that the spreading of Europeans in all continents – a process started long before the Industrial Revolution, but which gained momentum after the beginning of it – has taken place primarily through ocean navigation at the highest possible technological level. This obviously led to coastal concentrations in both countries of departure and territories of destination. In most cases, in the first stages of the process, the presence of Europeans overseas was limited to coastal areas and to such maritime powers as Great Britain, Holland, and France who followed this pattern of colonization. It was clear that whoever controlled the coastline had an important grip, not only on trading, but also on such practices as piracy, slavery, and smuggling. Only in the 19th century, when industrial technology was able to provide an efficient network of terrestrial transportation, did the Europeans succeed in getting a firm hold on territories far inland. Still, while further advances were in progress, coastal areas remained of the utmost importance to the whole expanding system, as it was through ports, coastal cities, river outlets, and other infrastructures, that the flow of expansion was regulated. Such port-cities as New York, Rio de Janeiro, Rotterdam, Singapore, Casablanca, Shanghai, and Sydney, not only played a major role as nodes of this system, but successively became poles of huge littoral agglomerations.

Throughout the Modern Age and into the Contemporary, it became clear that the spatial relation network established in ancient times in the Mediterranean was extensible to the whole world; and that the oceans would become the fundamental media for trading and conquest and long-range migration as long as they could be dominated by navigational devices deriving from industrial technology. The links between these new views and coastal concentration look too obvious to need further discussion. They confirm and accentuate the pre-industrial trend between the density of coastal settlement and the intensity of the push for the use of the sea, however motivated.

Cultural changes related to the transition from pre-industrial to industrial societies have also played a major role in encouraging migrations towards the coast. Inasmuch as areas of sea/land interface – with better housing accommodation, a wider range of facilities, easier communications, more economic opportunities, and a milder climate – provide improved standards of living and richer social relations, they are apt to be places of immigration. In general, the demand for more diversified relations and opportunities has been a leading factor for the agglomerative process that gave way to the urban age. Coastal areas, with their typically linear features, have certainly exerted a strong attraction, resulting in high densities of settlement and population.

Agriculture – or rather industrial agriculture – and the need for acquiring new land to meet overpopulation problems, has been another important factor of coastal clustering, especially in terms of new settlements. Up to the last century, vast expanses of lowlands were permanently or temporarily flooded and their reclamation was only possible through artificial filling and/or the construction of sea walls; the easier way of establishing a drainage system through the digging of canals could not always be adopted on account of low elevations. The actual lifting of water through pumping could only be accomplished when the amount of energy needed was made available in the latter part of the 1800s. Not only did these operations lead to the acquisition of new farmland and the establishment of rural coastal settlements, but they also provided vast areas of land to meet the uprising demand for tourism and recreation, which developed as a by-product of the industrial culture.

This demand has undoubtedly been one of the main factors in determining the coastal concentration of humans over the present century and has led to a wide range of processes and consequences, which will be specifically investigated in this book through the presentation of a number of case studies. Before going further, however, two basic questions seem to deserve further preliminary discussion. Why are coastal areas privileged places for recreational uses? Or, which factors, *specific* to the coastal zone, encourage these uses? The second question is: what *specific* effects do recreational uses originate in the coastal zone?

A temptative answer to the first question should move from considerations as to the origins of leisure time and the practice of modern recreation.

Why recreation in the coastal zone?

The concept of leisure has been discussed in terms of time, activities, and attitudes of mind. 'Leisure can be regarded as a measure of time: it is the time remaining after work, ... and available for doing as one chooses. It may be defined as 'discretionary time'.'[4] As to recreation, 'it embraces the wide variety of activities which are undertaken during leisure'.[5] In other words, 'any activity of leisure time undertaken by choice and for pleasure would constitute recreation.'[6]

If we are to accept these definitions – which seem reasonable – leisure and recreation at large are obviously as old as mankind: their equivalents in different languages may, in fact, be traced in the oldest written documents, the Bible being the best known example. However, the concepts and the terminology used for it do change over time and the attribute of *modern* which we have used above in connection with both leisure and recreation, adds and substracts something to these concepts.

One leisure day out of every lunar phase has long been established through various religious rules as they began to influence and guide social behaviour and everyday life: the Christian Sunday and the Jewish Sabbath are examples. The original meaning of these non-working days, however, was to organize social time not so much for leisure as for religious practices. On the other hand, pre-industrial societies were not familiar with a clear discrimination between working time and leisure time and there was nothing in their customs resembling a modern working schedule. The agricultural labour depended – and largely depends today – on such variables as the time of the year, the weather, and the necessities related to crops and domestication. In such situations, leisure results if and when possible, and for non-working people it could actually extend to the whole lifetime.

On the contrary, *modern* leisure – and thus *modern* recreation – is the result of a mass discrimination in the organization of time between working and non-working hours. This has implied the perception of time *by the clock* instead of *by the sun* and was the result of the industrial organization of labour, as imposed by the new needs of mass production. This organizational model extended to other non-industrial activities, including agriculture itself, but it started in the factory and may be regarded as a fundamental aspect of the Industrial Revolution. The 60 hours (or more) per week working schedule commonly adopted in early industrial times, certainly did not allow for much leisure; nevertheless, that was precisely the start of what has been referred to as modern leisure. Because it gave potential space for leisure in the evenings (which came to be more and more artificially lit, thus multiplying the opportunities for recreation) and on Sundays, through a shrinkage of religious practices and a 'contamination' of them with recreational activities. An expanding social framework (from the farmhouse to the factory, from the village to the city) also stimulated social activities which were partly of a recreational nature. From a social standpoint, however, one of the main products of emerging industrial societies has notoriously been the formation of an urban middle-class, provided with more leisure time, higher revenues, better opportunities and facilities, a sharper curiosity, and a more acute inclination to be *à la mode* or, in other words, to do what others do.

The original cultural setting of the 'recreational revolution' was early Victorian England: the same setting in which an industrial society took over for the first time from a rural one. 'Certainly the middle-class young enjoyed more free time than their elders had done, for the increasing emphasis upon public schools and university education as requirements for gentility meant a prolonged freedom from the immediate pressure of earning a living,' writes P. Bailey about Britain in the mid-1800s.[7] And a gazetteer of the time records that 'the cheap press ... has transformed the severely domesticated Briton into an eager, actively inquiring, socially omniscient citizen of the world, ever on the outlook for new excitements, habitually demanding social pleasure in fresh forms.'[8]

This Victorian England provided the same setting where *tourism* originated: as a form of recreation as well as a neologism. Travelling in itself is, of course, as old as mankind, including travelling for leisure, or curiosity or other such purposes regarded as recreational. But the kind of travelling which upper and middle classes started in mid-19th century Britain was something unprecedented and it was so perceived as to require a new word to refer to it. Tourism was – and essentially still is – recreational travelling, extending from individuals to social groups. Not just a cultural fact, as it had become with increasing evidence over the previous two centuries, but a social process, implying the development of specific news structures (the 'tourist offer') and the availability of means of transportation which, once again, were products of the Industrial Revolution. 'The "habit of enjoyment" was in fact diffused and encouraged through major improvements in communications', refers Bailey again.[9] 'By the early 1850s, the major lines in the British rail system were completed or under construction. Rail travel stimulated a general public curiosity and helped break down regional insularities of mind and practice ... The railway gave a new mobility in leisure and the regular spate of advice and reports in the press in the summer months testified to the growing habit and ritual of the annual holiday.'

A demand for therapy and body fitness, efficiently stimulated by the growing influence of the medical class, had been, in earlier times, a major factor for moving to such health resorts as spas and other watering places. The spas might offer 'but London life on another stage', as Elizabeth Montagu had written back in 1754,[10] 'but the change of environment they provided was as important as the pleasure they could afford'. However, in the course of the following century, the privileges of the few came within grasp of the many. While urban growth related to industrialization was accompanied by a deterioration of the urban environment, industrialization itself resulted in an increase of both wealth and leisure time and multiplied the possibilities for mobility. These new opportunities led in Britain, and successively in other industrializing countries, to the 'rush to the seaside' and to the establishing of new resorts along the waterfront.

These considerations on the origins of coastal recreation take us back to our first question. Are there *specific* factors which privilege coastal areas for recreational uses and vacationing? The appeal of the sea – of its uses, but of its sight and perception as well – to the British is commonplace; but it is true and deeply felt. This may partially explain the case of Britain and other countries with attitudes deeply rooted in the relationships with the sea. But it is not the general case and other possible factors should be investigated. Among these, the most widely quoted seem related with such peculiarities of the coast as the climate, the scenery, and the quality of the air. Also, the coastal zone provides the possibility of practising a wide range of recreational activities – swimming, surfing, sailing, boating, fishing, diving, sunbathing – which would be impossible, or partially inhibited, or not as satisfactory elsewhere.

The sum of these facts does give part of an explanation as to the 'coastal appeal'; but not all of it. The countryside also offers in general better environmental qualities and, in many areas, wider open-door recreational opportunities, as compared to urban areas; but its appeal to urban dwellers has always been largely inferior when compared to the seaside. Mountain areas may be credited – again in a general sense – with offering an environmental setting and recreational opportunities not inferior to coastal areas. They do not, however, enjoy as high a demand for vacation and recreation as rivieras. In conclusion, environmental qualities and recreational opportunities do not seem to provide sound enough reasons to explain why the demand for recreation still ranks higher for coastal places than for any other. A more complete and satisfactory answer to the question could perhaps result from considering other facts, originating from economic trends and from social psychology and framed in a spatial and historical background.

It should be firstly considered that quite often in market-oriented economies, the demand tends to be a response to the offer. The more pressing the offer, the higher the demand, and there is no doubt that the offer of recreation and/or vacation from coastal areas exceeds by far, both in intensity and variety, offers from any other place. The response to this offer is a higher demand and once this process is started, it gains momentum through the socially diffused trend of imitation: people follow other people. These considerations, however, shift the problem to another question: why is the offer of coastal recreation higher? The answer needs more investigation and only a hypothetical one can be given here. In territorial assets resulting from the diffusion of industrial cultures, coastal areas tend to be 'strong' areas, i.e. areas where people, capitals and economic opportunities have tended to concentrate as a result of the processes discussed above and which have been progressing independently – at least partially – from the demand for recreation. On the contrary, other non-urban areas, and more so mountain areas, have been 'weakened' by a migration process, which, again, has involved not only people, but also economic synergies. These are general trends of course, but the numerous exceptions do confirm the rule: strong mountain regions do exert a high offer for recreation/vacation and get a high response in terms of attendance; on the other hand, weak and depopulated coastlands exert a low offer and get a low response, whatever the quality of the environment. The numerous cases of strong offers from weak areas again are no exception to the rule: in these cases, the strength of the offer is the result of enterprises and investments originating from the outside, and specifically from a strong industrial or urban area.

In conclusion, the logical sequence of the process seems to be: economically strong area → strong offer of vacation/recreation → strong response in terms of attendance and demand. Environmental factors do play an important role, but not a decisive one: vast and accessible regions with fine sceneries and high environmental standards are not used for recreation because they are not being marketed (and not necessarily as a result of conservationist policies); on the contrary, saturated resorts with environmental standards lower than in most cities keep being objects of high demand because they succeed in exerting high and qualified offers.

Last, but not least, the specific spatial setting of coastal areas should be taken into account. Unlike other regions which spread in a bi-dimensional space, coastal areas are essentially linear: they certainly have a width, but this second dimension is irrelevant when compared with the alongshore dimension. In fact, the spatial organization of a coastal area hinges upon a line and rapidly loses significance within a short distance from it. The location of human activities in a bi-dimensional space can be ubiquitous or at least flexible: activities to be located along coast, inasmuch as they are related to the land/sea interface, cannot have the same degree of flexibility. They are bound to locate at the closest possible distance from the coastline and this rigidity implies concentration. A tourist settlement in a mountain area may be theoretically expanded in all directions; a coastal recreational area can only be developed along the waterfront or at a close distance from it. This again calls for higher concentration, higher costs, and a more pressing offer on the market, to cover these costs.

These considerations also suggest some explanation as to our second question, regarding *specific* effects of recreational uses on coastal areas. Such effects as housing development, saturation, environmental degradation and changes in the cultural and economic frames look rather obvious, as well as unspecific. Some of these however, may be emphasized by such coastal peculiarities as the uni-dimensional setting, the accentuated multi-use conflict, or the fact that the coastline is the place where all water-carried wastes tend to concentrate.

Inman and Brush[11] suggest that human presence along coasts takes three interrelated forms: a) the impact deriving from high densities and multi-use, b) environmental pollution, with particular emphasis on waters, and c) the critical modification of natural balances both in the ecology of living organisms and in the sources of sediments that constitute beaches. Such negative consequences tend to emphasize as a result of technological development not being met by adequate management and planning policies.

In an attempt to capitalize on the resources of the seashore, man's vision is limited by the tricky assumption that technology can do almost anything. This is true to a certain extent. But the medals produced by technology for man's benefit have their reverses. The demand on natural mineral deposits has decreased as a result of a massive replacement of metals with synthetic plastics and this has been limiting deleterious environmental effects due to mining activities; but plastics decay very slowly and discarded containers are piling up in quantities which will pollute the beaches for centuries to come. Improved transportation has been one major factor for tourism and has increased the supply of land and water available for recreation; however it has also been a major factor for the diffusion of man and pollution. In fact, and in spite of any technology, human pressure inevitably leads to ecological disturbances, alteration, and artificialization of the coastline, with the final result of a reduction in the degree of attractiveness of the resource.

For a number of years, the practice of planning was thought to be the best antidote against human concentration and multi-use conflicts, and the only available referee in the match technology vs. environment. However, many years of failure have opened the eyes of many optimistic viewers. Planning is the tool of the public, its weapon against private interests: but not always – in fact, in very few cases – does the public prove to be the stronger part. In many countries, planning is inadequate; in most others, it is just speculation. Most negative effects of tourist development came from poor planning or no planning, and what had been a major process in raising social and economic conditions in underdeveloped coastal regions, turned, in many a case, to destroying the sources of its own success.

These concepts have been stressed by numerous authors over the last fifteen years and it sounds superfluous to insist on them here. They have been recalled, however, as they constitute a rather common framework for the case studies presented in this book, providing numerous and worldwide examples of the deterioration of natural ecosystems as a consequence of high technology and unplanned development for recreation.

The 23 contributions have been grouped into four sections, according to what has been regarded as the main approach of each. The introducing section includes eight regionally-based studies, discussing recreation in such coastal areas as Florida, South-Wales, Port Phillip Bay (Southern Australia), Singapore, the Azov Sea coast, Guyana, Central Chile and Québec. Not only are these areas scattered over both hemispheres and all five continents, but they present an extremely wide range of situations in climate and environmental assets, as well as in historical and cultural backgrounds. There are similarities, however, in the nature of their problems, and in the way these are managed (or mismanaged), which show a substantially uniform trend in man's recreational impact with the coastal zone.

Contributions grouped in the second section show that this impact is not necessarily related with mild climate, sunny conditions, or the presence of beaches. With the only condition being the presence of people enjoying certain economic and technological standards, the demand for coastal recreation is diffused worldwide, regardless of apparently adverse environmental conditions: this is the case of such areas as the State of Washington, U.S.A., the nearby Pacific waterfront of Canada, and the fjords of Norway, where harsh climatic conditions and low water temperatures may inhibit certain activities, but do encourage others. This is also the case of the semi-arid and poorly accessible Gulf coast of Mexico, of the beachless and subsiding coasts of Louisiana, and of the western coast of the Caspian Sea, where the sea level has risen by over 3 m during the last half century.

Section III includes seven case studies on coastal planning in general. On global planning, as in the case of Köge Bay, near Copenhagen, or of the coast of Aquitaine, close to Bordeaux, where the pressures of large urban areas on sandy beaches has been particularly strong, and also on sectorial planning, as in the management of surfing along Sydney's southern beaches or in the construction of artificial structures along stretches of Israel's eroding coastline. The impact on beach erosion of recreational use intensity is also considered in a detailed study of a limited area along Lake Ontario, while a most interesting experiment is described on the effects on a beach profile and its grain size of people walking on it. Finally, the results of an investigation conducted through a questionnaire and directed to users of the Biscayne National Park (Florida) are presented. As the title indicates, this was aimed at formulating management policies through the use of the visitor's perception and may certainly be regarded as a planning operation.

The last two essays, which form Section IV, could not be included in any of the previous sections, on account of their unique approaches. The first one, which is also the longest of all, provides an overall picture on marine recreation in North America, moving from its historical and cultural antecedents and next considering its evolution in relation to technological advancements, the various kinds of demands, and the land-based supply system, which, in many cases, turns out to be an important factor in orienting coastal economies. The very interesting closing paper illustrates morphological and functional models of beach resorts through the application of semiotics and supplies an original approach to the understanding of the morphology of many coastal resorts.

As a whole, these contributions enforce the belief that coastal zones are key areas, where careful land use controls and sound planning measures need implementation, not as much to the benefit of tourist industry, but more so to the well-being of man on Earth. No detailed calculation will be needed to demonstrate the increasing value of these resources. This will find expression in the actions of the ordinary people as they talk about the resource and seek to use them. New demands generate new problems. For better or worse, we live in an age where the 'natural' environment is largely what we make it. We can attain most of what we want if we really try; and if we fail, the fault is ours.

Paolo Fabbri
Department of Historical Disciplines,
University of Bologna,
Italy

References

1. G. W. F. Hegel, 1955, *Vorlesungen über die Philosophie der Weltgeschichte*, J. Hoffmeister, I p. 197.
2. C. F. Bennett, 1975, *Man and Earth's Ecosystems*, J. Wiley, p. 291.

3. P. Gourou, 1973, *Pour une géographie humaine*, Flammarion, Ch. X.
4. A. Mathieson and G. Wall, 1982, *Tourism. Economic, Physical and Social Impacts*, Longman, p. 7.
5. *Ibid.*
6. A. Phelps, 1988, *Seasonality in Tourism and Recreation* ..., in 'Leisure Studies' Vol. VII No. 1, p. 34.
7. P. Bailey, 1978, *Leisure and Class in Victorian England*, Routledge and Kegan Paul, p. 87.
8. J. H. S. Escott, 1897, *Social Transformations of the Victorian Age*, London, p. 14.
9. Reference 7, p. 59.
10. J. A. Patmore, 1961, *The Spa Towns in Britain*, in 'Urbanization and its Problems', Blackwell, p. 32.
11. D. L. Inman and B. M. Brush, 1973, *The Coastal Challenge*, Science **181**, pp. 20–32.

Acknowledgements

The collection of such an unprecedented array of studies on the topic of coastal recreation has not been an easy task for the editor. Hundreds of letters were mailed and received over a period of three years and many dicussions were held at meetings and on the field. Everything, however, was greatly facilitated by the editor's opportunity of serving as a full member of such a stimulating group as the Commission on the Coastal Environment of the International Geographical Union. Under the friendly and enthusiastic guidance of Roland Paskoff, the Commission has given full support to this project and has proved to be the best possible medium – if not the only one – for improving international cooperation in this field of study. There is a diffuse belief that the CCE only deals with coastal geomorphology. This book, wholly produced within the Commission, proves that it is not so.

Many people should be acknowledged: primarily the contributors themselves, among whom so many have become so close to the editor's feelings as to establish precious ties of friendship. Other people and institutions, however, should be mentioned, in an informal and incomplete list of acknowledgements. Jess Walker, Norb Psuty, and Doug Elvers have been generous with their advice and understanding. Wolf Tietze, editor-in-chief of *Geojournal*, promptly appreciated the value of the project and gave decisive support for publication in this series. Kluwer Academic Publishers have bravely undertaken all the publication and distribution processes. The University of Bologna, having just celebrated its 900th birthday, has provided a financial contribution for publication. Thanks again to everybody and ... let's keep going!

SECTION 1

Regional Studies

1. The recreational use and abuse of the coastline of Florida

Introduction

The State of Florida possesses one of the world's most heavily developed recreational shorelines. Much of this development has taken place on low barrier island coasts, well known for their morphological and ecological fragility. The history of recreational development in Florida has been largely interventionist, geared to speculative demand and quick profit, without due regard for the environment. In the mid-1960s it became clear that development could not continue unchecked. The environment, particularly in southern Florida, showed all the symptoms of stress, and even the business community were concerned lest 'pollution' should lead to falling demand (Carter, 1974; Blake, 1980).

In many respects, Florida's attempts to balance recreational potential with shoreline environment provide us with a seminal experience, the lessons of which should ease the burdens for coastal managers worldwide. This paper explores some of the geographical background to the 'crisis in paradise'.

Florida: the geographical setting

Florida comprises a low-lying peninsula (140 212 km^2) in the southeast corner of the USA (Figure 1), dividing the Gulf of Mexico from the Atlantic Ocean. The most prominent physiographic feature of Florida is, paradoxically, its extreme flatness. Only 10% of the State exceeds 80 m above sea-level, and the highest point is 112 m. This flat, southern peninsula is formed of geologically-young Cenozoic rocks (mainly Oligocene and Miocene argillaceous limestones), and more recent (Pleistocene/Holocene) estuarine, eolian and shallow marine deposits. Inevitably, there is a strong interaction between the land and the sea, especially in terms of climate, biota, and human activities.

The coastline of Florida is 1900 km long; with the exception of Alaska, the longest of any US state. The coast is composed of barrier islands, beach ridge plains, lagoons, deltas, reefs, cheniers, estuaries, bayous, saltmarshes and mangals. There are major contrasts between Gulf and Atlantic coasts in terms of wave energy, ecology and land use (see Figure 1). The energy contrasts are important in longshore drifting of sediment, nutrient recycling and scenic beauty, but average conditions belie the important role of extreme storms, especially tropical hurricanes, in shaping the Florida shoreline and its activities.

A combination of natural attractiveness and entrepreneurial skill have been the twin driving forces behind the tourist 'boom' in Florida. The State attracts around 32 million visitors a year, contributing (1980 statistics) $1400 billion to the

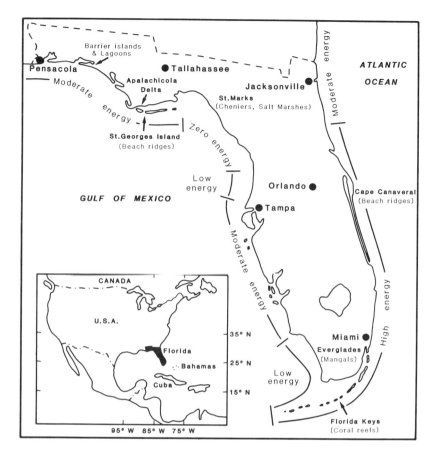

Figure 1. Geographical setting of Florida.

economy. To service this burgeoning industry, Florida's population has risen spectacularly since 1950, increasing by 40% in the decade from 1970 alone. Socio-economic and ethnic profiles of incoming migrants are interesting, including extremes of both very rich and very poor, religiously persecuted, politically oppressed and criminally intentioned, from both Europe and Central America, as well as high numbers of elderly North Americans. Many of these groups gravitate towards existing ethnic or class concentrations (or ghettos), so that certain areas have become dominated by particular groups; rich Jewish widows in Miami Beach ('Wrinkle City'), Cubans in south Miami, elderly in Collier County and so forth. Such cohorts raise additional problems for coastal managers, particularly in terms of communication and education. One pertinent example is the large number of elderly retirees living on the coast, as many have no appreciation of the threat posed by coastal storms and are often reluctant to evacuate if the need arises.

The impacts of recreational development on the coast of Florida are numerous. Most impacts stem from the pervasive and seemingly insatiable appetite for coastal land. Much of the land reallocation is purely speculative, founded on the State's long time reputation as a development 'booster'. Recreation development has destroyed beaches, dunes, reefs, lagoons and wetlands, changed access and above all attracted people. These activities have, in turn, perturbed or upset the balance of nature, both encouraging and destroying biotic communities, disrupting sediment transport, polluting coastal waters and exploiting resources, like groundwater, shells and corals, fish and birds. To appreciate these processes it is necessary to outline, briefly, the history of coastal development in Florida.

Coastal development history of Florida

At the time of Statehood in 1845, most of the population of Florida was confined to the north of the state between Jacksonville and Pensacola. The coast was largely uninhabited, dangerous to sailors and of little direct use to settlers.

Coastal development started in the 1880s with the spread of the railroad system into peninsular Florida. Prominent among the railroad 'barons' was Henry Flager of Standard Oil, who by constructing and judicious purchasing extended his Florida East Coast Railroad south from Jacksonville, through Daytona, Palm Beach, Miami and even as far as Key West, acquiring some 470 000 ha of 'development' land from the Federal Government in the process. In addition to exporting agricultural goods and minerals to the northeast, the railway brought tourists.

Throughout the State, Flager and others were active in 'improving' the land, through drainage and clearing of natural vegetation, allowing access and settlement. These environmentally largely deleterious activities were not only condoned by the State, but encouraged through generous Land Grant Acts, ideal for avaricious speculators.

Early tourists were attracted by the wilderness of interior Florida and its potential for hunting, as well as by the attractiveness of the beaches and the winter climate. As visitors to Florida increased so pressures on oceanfront land rose, especially from the property owning, overwintering northeasterners, clustering in exclusive seaside enclaves, like Palm Beach, Winter Park and Sarasota. Rising demand for shorefront property led to the destruction of much of the natural coastline, including the levelling of dunes, dredging of lagoons and reefs and the infilling of back barrier bays. In places artificial 'cuts' were made through barrier islands to link the bays with the ocean, examples include Bakers Haulover at Bal Harbour and Sikes Cut on St. George Island. After 1900, settlements began to spread out along the barrier islands, especially in Dade and Brevard Counties in the extreme southeast and in Hillsborough and Pinellas Counties on the central Gulf coast around Tampa, St. Petersburg and Clearwater.

The origins and early development of Miami Beach highlight many of Florida's growing pains. Around the turn of the twentieth century, a number of swampland scandals broke (Carter 1974, p. 69 et seq), in which undrained, uninhabitable land was being sold to out-of-state purchasers as 'Promised Land' or 'America's Tropical Paradise'. (This gimmick has been repeated at intervals ever since, to a seemingly bottomless pool of gullible customers). Attention soon focussed

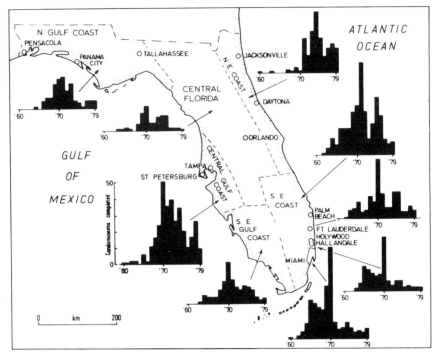

Figure 2. Condominium growth patterns between 1960 and 1979. Construction as shifted from the southeast to the northwest. The slump in the middle 1970s is partly due to implementation of legislative control.

on the mangrove and sand dune coast around the embryonic settlement of Miami. In 1910, an Indiana oil millionaire, Cal Fisher, arrived in Miami and joined forces with a local farmer, John Collins who owned 8 km of barrier island. As partners the two men infilled and reclaimed large areas of back barrier swamp, obtaining permits from both the Internal Improvements Board of the State Legislature and the US Army Corps of Engineers (the latter would only withhold permission if navigation was threatened). The wholesale destruction of the natural environment not only created new land, but eradicated many 'irritations' like alligators and mosquitos. However, the investment gamble of Fisher and Collins paid off in spectacular fashion, as they sold over $100 million of real estate, and prompted many others to follow their lead and 'reclaim' the lagoons and barriers of Florida.

Consolidation and expansion of coastal communities in Florida continued unabashed throughout the first half of the twentieth century. Declining fortunes of the railroads, were compensated for, and in no small measure, by the growth in car ownership and the extension of the state and the interstate highways, and more recently the spread of air traffic. Growth of the latter has been phenomenal: in 1981 there were 134 licensed airports in Florida, 15 000 aeroplanes and over 10 million take-offs and landings every year. Deregulation of commercial airlines since 1983 has added more to this total. The main result of these changes has been an opening-

Figure 3. Condominiums lining the shore at Daytona Beach.

up of Florida to an international market, and the penetration of less-developed parts of the States' coastline, particularly along the northern Gulf Coast.

The sale of Florida's coast has continued up to the present day, although geographical focus has shifted from the southeast to the northwest. Growth since 1960 has been largely associated with condominiums (Figures 2 and 3). These residential communities (increasingly time-shared), provide not only individual dwelling units, but also recreation facilities (swimming pools, handball and tennis courts, occasionally golf courses) plus various services (security, medical and catering). Between 1960 and 1980 over 1500 condominiums were built in Florida (Carter, 1982), providing accommodation for 750 000 people. There are many inconsistencies about the 'condo' boom, which has gained a somewhat tarnished reputation, redolant of the earlier swamp land scandals.

Since 1970, Florida has been marketed internationally, partly as a tactic to overcome the US offseason (March to November) by matching it with the European holiday period. (Few Europeans appreciate how hot and humid Florida is in summer.) By 1978, 300 000 foreigners were visiting Florida, but variable exchange rates, plus poor publicity, especially in respect to violent crime, have made it a volatile market. Efforts have been made to promote Florida as an international 'playground', mainly through the opening of various theme parks – Epcot, Sea World, Circus World – or leisure parks – Disney World and Wet 'n' Wild. While most of these leisure parks are not coastal, most patrons visit the beach at some time during their trip.

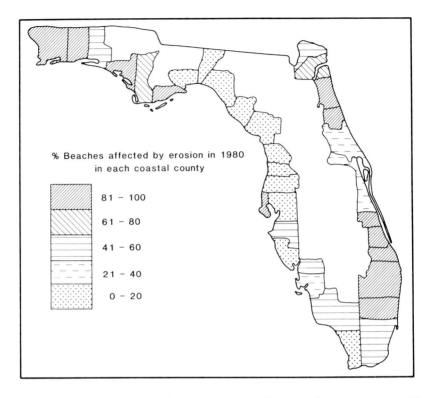

Figure 4. Shoreline erosion in Florida correlates with recreation development. Man's enthusiasm to be beside the sea is often damped by the onset of erosion.

Environmental impact of coastal recreation

Shoreline recreational developments

On all US coasts there has been a tendency to build as close to the shoreline as possible; Florida is no exception. Such actions have destroyed dunes, wetlands and beaches which formed natural protective barriers against storms and floods. To the developer the waterfront, maximises profit; to the owner it provides access and unimpeded views. Yet, over time, such locations bear the brunt of natural disasters, and the risks of property damage and even loss of life are significant.

The first shoreline buildings were beach houses in the dunes. Very often the seawardmost dunes were lowered or removed altogether to give a view of the sea. Very soon house owners became aware of shoreline changes, expecially natural erosion, and began to protect against it. Much of this protection was unapproved, unsightly and ineffective. Along the east coast, bulkheads and groynes were common after 1925, yet by the mid 1930s much of the dune line was destroyed (Hansen, 1947). Initially, State or Federal intervention in beach erosion was negligible, despite rising nationwide concern (Quinn, 1977). Money for erosion

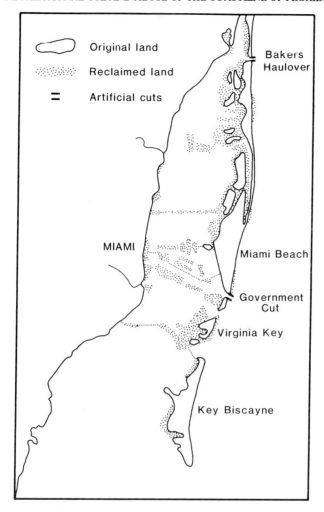

Figure 5. Location map of Miami Beach area.

control projects was limited, and work was piecemeal and uncoordinated, even between neighbours. It quickly became clear that such an approach was exacerbating erosion, and there was mounting pressure for official assistance. To some extent, Florida became a natural laboratory for shore protection devices, including inlet by-passing and back-passing, beach nourishment and diverse species of revetments, breakwaters and groynes (Bruun and Manohar, 1963). Not all were successful.

The Florida Coast Management Program (1981) has identified about 870 km of eroding shore (Figure 4), of which about 40% is considered 'critical', threatening roads, houses or other structures. Most of the problems are on the upper and lower east coasts, and on the Panhandle Coast, west of Appalachicola. Worst erosion coincides with the main recreation areas; where there is no recreation, for example, on the 'Big Bend' coast north of Clearwater, no erosion is recorded.

In 1970, Florida pioneered the concept of 'set-back' lines for coastal developments (Purpura, 1972), whereby a line is drawn parallel to the shore, seaward of which development is prohibited. The original set-back lines were 25 or 50 feet (8 or 17 m) from the dune line of HWM, but more recent lines are based on local conditions and may be 100 m or more inland. Before a development control line is established a public hearing is held, after which construction can only proceed if a permit is issued by the State. The number of permits has risen, since 1970, to over 700 per year (Balsillie et al. 1983). Failure to obtain a permit can result in a $10 000 fine and an order to remove unauthorised structures.

Miami Beach – case study. Miami Beach (Figure 5) is one of the most valuable coastal resorts in the world. Real estate alone is probably worth in excess of $800 million. However, much of the resort was built on a shifting barrier island, and subsequent development highlighted this dynamic quality, and led to a serious shoreline erosion problem, threatening not only the shoreline hotels and condominiums, but also the tourist and recreation industry itself. No beach equals no visitors. A number of factors are responsible for the severe erosion problem. One, properties were built too close to the sea. Two, ineffective and disorganised protection works were implemented based on little or no appreciation of natural processes. Three, artificial inlets, excavated across the barrier island – Bakers Haulover and Government Cut – impeded longshore drift. Four, a lack of procedural and fiscal means to tackle the problems properly. The early construction of bulkheads and groynes was effective inasmuch as it fixed the shoreline, but equally it was instrumental in destroying the beach. Between 1926 and 1960 the shoreface steepened and migrated inshore at a rate of 2 m/y, resulting in a narrowing of the beach. Disruption of longshore drift by jetties, inlets and groynes may have been partly to blame for this sediment starvation, certainly it was the focus of numerous legal disputes (Mills et al. 1976). In recent years, beach nourishment has been widely practised, although even this has not been wholly successful – see Carter (1987). The largest scheme has seen the virtual reconstruction of 13.5 km of beach at a cost of over $80 million.

Summary. From the early protection of individual shorefront properties, Florida has entered a spiral of escalating problems and costs. Today the strategy combines both prevention of shoreline development and protection of existing structures, through a mix of engineering, ecological and planning techniques. Funding for erosion control can come from City, County, State or Federal sources, in some instances via taxes or bond issues. The State operates an 'Erosion Control Fund Trust Account' for disbursement of aid where erosion is severe, although to qualify individuals and communities, certain environmental impact standards and access provisions must be met.

Access to Florida's shoreline

As Graber (1981) points out, Florida has been slow at expanding public access to the shore. It can be particularly galling for out-of-state vacationers to find that getting the last 50 metrs. to the beach presents the biggest challenge. However, improved shore access is a federal objective, enshrined in the Coastal Zone Management Act (1972). In 1976, the Florida Department of Natural Resources (DNR) estimated that 77% of shorefront was privately owned, and although the public held rights to the foreshore by custom, prescription or under the public trust doctrine, gaining access was not easy (Maloney *et al.*, 1977). Clamour from private owners for public assistance in shore management has enabled the State to trade help for access rights. In addition, federal funds are available to acquire access. In 1983, the Florida legislature started a 'Save our Coast' programme (Glassman, 1983), not only to conserve and protect valuable coastlands, but also to secure access. One noticeable manifestation of this policy has been the development of coastal parks; at present there are six National Seashores and 24 State Parks in Florida, satisfying recreational demands ranging from camping to fishing. In all parks an entrance fee is charged, and the extent of public use managed within the notional natural carrying capacity of the environment. It would be fallacious to conclude that all beach access is difficult; in many places like Panama City Beach, Ormonde-by-the-Sea and Daytona public entry to the beach is encouraged, and often parking is allowed between the tidemarks. The new Miami Beach Boardwalk has transformed a previously run-down and notoriously inaccessible stretch of coast, where beach access were all too often down alleyways among the garbage skips of hotels.

The publication in 1985 of 'Florida's Sandy Beaches: an Access Guide' (Fischer, 1985) has provided an invaluable source of information on how to get to the beach and spread your towel.

Coastal management for recreation

In common with almost all coastal states, Florida has prepared a Coastal Management Plan (FCMP) under the aegis of the Coastal Zone Management Act of 1972. While the FCMP provisions extend far beyond recreational aspects, almost all Florida's coastal activities are tied in some way, to the leisure industry.

The draft FCMP was presented in 1980 for public hearings and comment and the final version was approved in 1981. The Plan defines coastal boundaries, outlines the range and importance of coastal activities and covers the legal and decision-making structure for coastal affairs. A major objective of this, and all US coastal zone management plans prepared under the CZMA, is to achieve consistency, at both State and Federal levels, in terms of policies, decisions and standards.

The FCMP recognises the acute pressures facing much of the Florida coast, as competitions for space increases. Conflicts have arisen between the fishing, water abstraction, waste discharge, power plant siting and, inevitably, recreation. The development process has been curbed and directed towards specific zones. Local and State ordinances have been passed to restrict coastal development, often in the realisation that over-development can be inimical to continued prosperity.

Recognition that the Florida coastline represents a valuable recreational resource, the Florida Department of Natural Resources (DNR) has been empowered, since

Figure 6a. Residential canals fill back barrier lagoons around Miami Beach.

Figure 6b. Bakers Haulover inlet impedes longshore drift and leads to down coast erosion.

the Outdoor Recreation and Conservation Act of 1963, to establish State Parks, wilderness areas and wildlife sanctuaries, and to develop their recreational potential. Obviously, in many cases, this potential is relatively low, and restricted to more adventurous and active visitors who are prepared to eschew comfort for seclusion. Thus coastal wildernesses, like St. Joseph's Peninsula and St. Vincent's Island have been opened up to small numbers of visitors, usually under supervision of rangers. The State Park system caters for larger numbers, often with vehicles, but once again, pressure ceilings can be imposed through access restrictions, entrance fees or rotational closures. The Florida Recreation Trails Act of 1979 was designed to encourage and fund the provision of trails for cyclists, joggers, horse riders and canoeists through undeveloped lands.

Coastal management in Florida has to grapple with numerous conflicts relating to allocation and use of land and water (Blake, 1980) as well as competing activities. Two particular areas of conflict are discussed below.

Coral reef management. Southeast Florida includes the only coral reefs in the contiguous US. The coral reef ecosystem is of considerable recreational and touristic value, while at the same time being very sensitive to disturbance. In the last 150 years the reef zone has shrunk by over 50%, so that it is now largely confined to the area south of Miami and along the Florida Keys. The reefs are popular with fishermen and sightseers, yet are susceptible to boat damage from engine oil pollution, wakes and anchors, and indirectly from dredging, waste water disposal and storm run-off. Reef die-back can have an adverse effect on inshore wave climates and biological productivity. Management of marine reef reserves has concentrated on five areas, which are managed for conservation, while still allowing some recreation, usually by electric or sail boats. There have been attempts to divert some activities, especially fishing, away from reef areas by providing artificial reefs.

Marines and residential canals. Boating has been a significant recreational activity in Florida dating from the steamboat era of the nineteenth century. Over 100 000 boats are licensed in Florida, many owned for recreational purposes, especially salt water fishing and cruising. To accommodate these vessels over 500 marinas have been built, plus many kilometres of residential canals (Figure 6a), cut or dredged through back barrier wetlands. In some places channels have been cut through the barrier beaches to improve access (Figure 6b), and almost all natural inlets have been 'improved' by dredging and jetties.

Extensive modification of back barrier lagoons has impaired tidal efficiencies and led to poor flushing and excessive retention times in many canals. Such conditions leads to anoxia and perhaps eutrophication. Added to this there are risks of chronic pollution, especially from boat engines, marine paints and septic tank seepages. It is necessary to apply strict design criteria to avoid these problems (Morris, 1981), yet more often than not these are ignored. All new developments require permits from the Corps of Engineers, while larger schemes must conform to regulations under the Florida Environmental Land and Water Management Act.

St. George Island – case study. The 42 km long St. George Island was, until a 6 km causeway and bridge opened in 1963, only approachable by boat. Since the bridge was opened, this remarkably beautiful Gulf Coast island (Figure 7) has

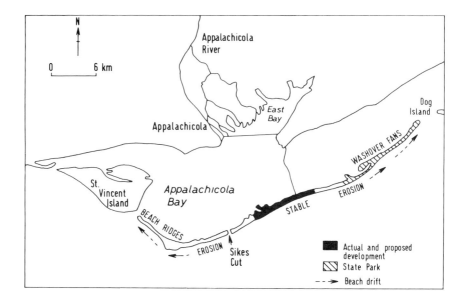

Figure 7. The location and land use of St. George Island, on the Gulf Coast of Florida.

suffered from developers and speculators. The island is divided between public and private owners. At the eastern end a State Park has been developed while to the west the land remains a wilderness area. In between, and around the bridge head, private developments are taking place (Figure 8). This land-owning pattern was partly the result of trade-offs between developers and State agencies. The developers faced many difficulties, not least of which were the provision of fresh water and the disposal of sewage (Livingstone, 1976). Notwithstanding, sales of lots, at prices ranging from $10 000 to $50 000 have been brisk. Although the developers are committed to environmental protection, it is clear that damage is increasing, with disturbance and clearing of vegetation and accumulation of fresh blown sand. Yet it is now widely appreciated that the most stable barrier islands are the most dynamic (Carter, 1988) so that efforts to 'fix' the shoreline may be misplaced.

Summary. Coastal management for recreation has many facets. In Florida, most management is geared towards sustaining and developing the tourist industry, which may not be environmentally expedient. In the last decade, efforts have been made to draw together many of the management strands, to produce a cohesive strategy, which will both fulfill federal requirements under the CZMA (and other environmental legislation), as well as satisfy business interests. This policy is not without its critics. Guy (1983) points out that the FCMP is 'more like a list of problems than solutions ...', and fails to provide both adequate safeguards against misapplication and reasonable integration with other state programmes in areas like transportation and forestry. One major shortcoming appears in coordinating and collating information, so that cumulative effects are often ignored, necessary cross-

Figure 8. The natural beauty of St. George Island (A) is being invaded by developers (B).

referencing not undertaken and monitoring ineffective. Guy feels that public participation is not encouraged, and until it is, Florida's coast management program will only 'limp along'.

Recreation hazards

Coastal recreation in Florida is vulnerable not only to economic recession, but to natural disasters. Building too close to the shore has its drawbacks, in Florida one of the biggest is the threat from storm damage. The coast of Florida is prone to tropical cyclones and hurricanes, between the months of August and November. These intensive, fast-moving, storms can cause catastrophic damage both from direct wave and wind attack, and from associated tidal surges and terrestrial flooding. Although the annual likelihood of a hurricane striking ranges from 1 year in 7 on the southeast coast and on the Florida Keys, to 1 year in 50 on the northeast coast (Dunn and Miller, 1960), in recent years most hurricane landfalls have been on the Gulf Coast, avoiding the highly popular east and west coast recreation areas. Indeed Miami, well within the high-risk zone, has not experienced a major hurricane landfall since 1926, just before the start of intensive development. However, in October 1987 Hurricane Floyd made a close approach. Although hurricane warnings systems have improved greatly in recent years (Simpson and Riedl, 1981; Carter, 1987), the sheer density of many at-risk coastal sites, makes mitigation difficult. Native Floridians are all too aware of the dangers of hurricanes, but many visitors (and new immigrants) are not. (It is noteworthy that much of the recent publicity for Florida's attractions is aimed at increasing tourism in the hurricane season!). So despite detailed evacuation plans, there is no guarantee that they will work smoothly. It is also clear that some sites cannot be evacuated entirely given the usual 12 or 24 hour warnings. The barrier island community of Sanibel and Captiva Islands has a self-imposed development 'cap' to obviate this type of problem. An alternative to evacuation is to strengthen buildings against hurricane damage and provide safe havens in community centres. The rebuilding of eroded beaches and dunes also serves to add protection against hurricane surges. Despite all these measures it is clear that a severe storm could cause immeasurable damage to Florida's recreation industry, not only directly through loss of life and property, but also indirectly denting a carefully refined image of carefree paradise.

A longer-term threat is related to world sea-level rise. If US Environmental Protection Agency prognostications are correct, then sea-level could rise between 0.5 and 3.5 metres in the next century (Barth and Titus, 1984). This could have severe ramifications for Florida's recreation industry, both in terms of land loss and through disruption of services (water supply and waste disposal) and communications. Of all the US States, Florida perhaps stands to lose most from projected sea-level rise in the twenty-first century.

Conclusions

Coastal recreation is central to the well-being of Florida's economic prosperity. However development of recreation potential has brought with it many unwanted environmental problems ranging from shore erosion to loss of ecological diversity. All too often these problems have been fuelled by a lack of effective State control, and a poor understanding of environmental processes. The future is uncertain. A better management framework is offered by the 1981 Coastal Management Plan, yet the problems remain and in many cases are set to get worse. How much longer the Sunshine State can bask in the glory of its recreational opportunities is conjectural.

References

Balsillie, J. H., Athos, E. E., Bean, H. N., Clark, R. R., and Ryder, L. L. (1983) Florida's program of beach and coast preservation. In Monday, J. (ed) *Preventing Coastal Flood Disasters*. Natural Hazards Research Center, Special Publication Number 7, 109–122.

Barth, M. C. and Titus, J. G. (eds) (1984) *Greenhouse effect and sea-level rise*. Van Nostrand Reinhold, New York.

Blake, N. M. (1980) *Land into water: water into land*. University Presses of Florida, Tallahassee.

Bruun, P. and Manohar, M. (1963) *Coastal protection for Florida development and design*. Florida Engineering and Industrial Experiment Station, Bulletin Number 113, Gainesville, Florida.

Carter, L. J. (1974) *The Florida Experience*. Resources for the Future/John Hopkins University, Baltimore.

Carter, R. W. G. (1982) Condominiums in Florida. *Geography* **54**, 41–43.

Carter, R. W. G. (1987) Man's response to sea-level change. In Devoy, R. J. (ed) *Sea Surface Studies*, Croom Helm, Beckenham Kent, 464–498.

Carter, R. W. G. (1988) *Coastal Environments*. Academic Press, London.

Dunn, G. E. and Miller, B. I. (1960) *Atlantic Hurricanes*. Louisiana State University Press, Baton Rouge.

Fischer, D. W. (ed) (1985) *Florida's sandy beaches: An access guide*. University of West Florida Press, Pensacola.

Florida Coastal Management Program (1981) Final Environmental Impact Statement. NOAA/Florida DER, Tallahassee.

Glassman, H. (1983) The State of Florida 'Save our coast program'. In Monday, J. (ed) *Preventing Coastal Flood Disasters*, Natural Hazards Research Centre. Special Publication Number 7, 82–83.

Graber, P. H. F. (1981) The Law of the Coast in a Clamshell: Part IV – The Florida approach. *Shore and Beach* **49** (3), 13–20.

Guy, W. E. Jr. (1983) Florida's Coastal Management Program: A critical analysis. *Coastal Zone Management* **11**, 219–248.

Hansen, I. (1947) *Beach erosion studies in Florida*. Florida Engineering and Industrial Experiment Station Bulletin, Number 16, Gainesville, Florida.

Livingstone, R. J. (1976) Environmental considerations and the management of barrier islands: St. George Island and the Appalachicola Bay bar. In Clark, J. (ed) *Barrier Islands and Beaches*. The Conservation Foundation, Washington, 86–102.

Maloney, F. E., Fernandez, D., Parrish, A. R. Jr., and Reinders, J. M. (1977) Public beach access: A guaranteed place to spread your towel. *University of Florida Law Review* **29**, 853–879.

Mills, J. L., Woodson, R. D., and Solomons, K. (1976) *Compilations of laws relating to Florida coastal zone management*, Vol. 1. Dept. Natural Resources, Tallahassee.

Morris, F. W. IV (1981) *Residential canals and canal networks: Design and evaluation*. Florida Sea Grant Report, 43, Gainesville, Fl.

Purpura, J. A. (1972) Establishment of a coastal setback line in Florida. *Proceedings of the 13th Conference on Coastal Engineering*, 1599–1616.

Quinn, M. R. (1977) *The history of the Beach Erosion Board*, US Army Corps of Engineers, 1930–1963. Miscellaneous Report 77–9, U.S. Army, Washington.

Simpson, R. H. and Riedl, H. (1981) *The Hurricane and its impact*. Blackwell, Oxford.

R. W. G. Carter
Environmental Studies,
University of Ulster,
Coleraine,
Northern Ireland

2. Management strategies for coastal conservation in South Wales, U.K.

1. Introduction

Various bodies both statutory and voluntary are concerned with coastal conservation. However such involvement is rarely exclusive and for most agencies and organisations coastal conservation is just a part of a wider environmental remit. For example, even a specific coastal designation such as Heritage Coasts represents but one aspect of the work covered by the Countryside Commission.

Wales is a minor U.K. tourist destination, with visitors tending to make for the three National Parks (Pembrokeshire, Snowdonia, Brecon Beacons) or to the Gower Peninsula, sited west of Swansea. The popular image of South Wales remains one of coal tips and pollution. But the days of Rhondda's glory are over and big industry has largely deserted the Welsh Valleys in favour of more coastal and motorway-accessible sites. The coal tips are now being regraded and though poverty remains, so do large areas of unspoilt countryside on the moors above the valleys and in coastal pockets. While for magnificence South Wales scenery cannot compete on equal terms with the more popular tourist areas to the North and West, they afford solitude and openness, in contrast to the gloom of cramped, run-down development within the valleys. So too does the coast where until recently development pressure has remained slight, though the growth of towns like Barry and Penarth in the coal boom of the late nineteenth century was little short of astounding. Those undeveloped areas both coastal and inland even if nationally insignificant remain cherished and well-used local features and if the recently completed M4 motorway is now the lifeline for South Wales, the coast and surrounding hills remains the lung.

Figures 1 and 2 illustrate the study area extent of which the most acclaimed part is Gower. West of Gower lies the county of Dyfed, much of whose coast falls within the Pembrokeshire National Park, while to the north the fringes of Brecon Beacons National Park extend into the County of Mid Glamorgan.

Immediately east of Gower is the city of Swansea for whose inhabitants Gower is an eminently accessible playground. Hand in hand with this heavy recreation demand is the impact of development pressure on the peninsula to accommodate Swansea's growing population. Swansea is fortunate in having retained a fine sea front although the sands of Swansea Bay belie the heavily polluted waters and the beach is largely spurned in favour of the coves of Gower.

The eastern side of Swansea is heavily industrialized, dominated by large chemical plant and Port Talbot Steel Works between which is squeezed the town of Port Talbot. The towering chimneys and steel works cooling towers form a dramatic backdrop to the dune systems of Margam and Kenfig Burrows. On Margam Burrows there is now a boating reservoir and in the hills to the north-east

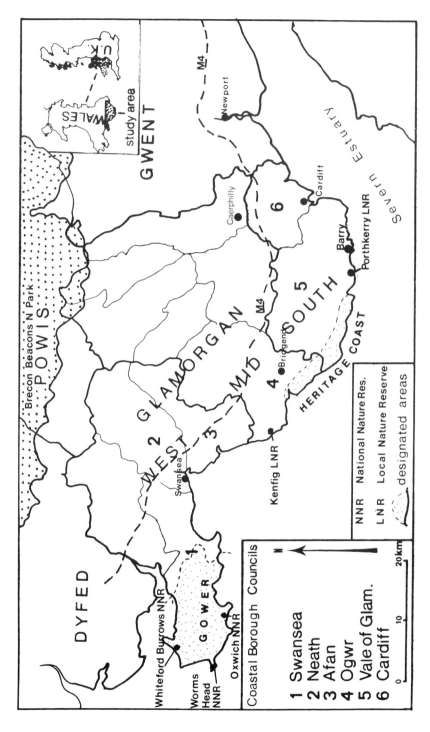

Figure 1. Location.

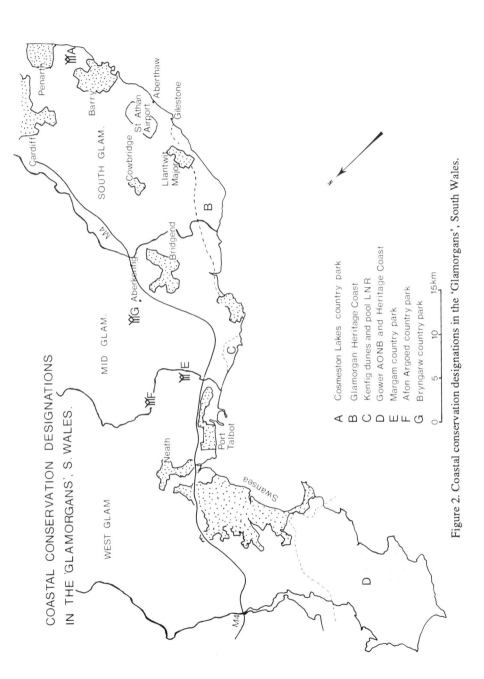

Figure 2. Coastal conservation designations in the 'Glamorgans', South Wales.

lie the extensive grounds of Margam Country Park (Figure 2). Adjoining Margam Burrows are Kenfig Burrows, the two being divided by the boundary between West and Mid Glamorgan (Figure 1). A Local Nature Reserve (LNR) has been established at Kenfig and surrounding golf courses provide a buffer between it and the resort town of Porthcawl. East of Porthcawl lies Newton Sands which mark the western end of the Glamorgan Heritage Coast and this extends 22 Km to Gileston. East of Gileston is the large power station and cement works of Aberthaw, Cardiff (Rhoose) Airport and further east still the town of Barry.

Following designation as a Rural Recreation Area, the coastline between Barry and Penarth has been made the subject of a Local Plan by the Vale of Glamorgan Borough Council (1980) and has been declared a coastal conservation area (Figure 2). This stretch of coast is bounded by a low line of cliffs which occasionally rises to over 45 m. Its geological significance is reflected in the (SSSI) designations of the greater part of Sully Island and also of the coast between Sawbridge Bay and Lower Penarth.

However the coast lacks the remote qualities of the Glamorgan Heritage Coast. Permanent caravan and chalet sites have been established on the cliff top which during peak periods accommodate up to 2000 visitors per day (Vale of Glamorgan Borough Council, 1980), and its proximity to Barry, Penarth and Cardiff makes it very much part of the urban fringe. In addition to the coastal frontage Cosmeston Lakes Country Park has been established some two miles inland on previously derelict land and the Local Plan provides for the management of this and adjoining open areas as part of the Rural Recreation Area.

2. Preamble

The Countryside Commission (CC) is responsible for conservation of scenery and provision for access and recreation in the countryside. In England and Wales National Parks and Areas of Outstanding National Beauty (AONBs) provide a statutory means of achieving these ends, although their success is questionable (McEwen, 1982). The powers of the Commission are restricted to the designation of boundaries for National Parks and AONBs subject to confirmation by the Secretary of State for the Environment. In its other roles the Commission acts in an advisory capacity, encouraging active conservation programmes with the carrot of grant-aid. Grant-aid is available to local authorities, voluntary bodies such as the National Trust and the private sector for a variety of specific activities ranging from tree-planting to the establishment of country parks and management of AONBs and Heritage Coasts. Most grants cover 50% of project costs though management agreements may be subject to a maximum grant of 75%.

Wildlife conservation is the responsibility of the Nature Conservancy Council (NCC) which like the Countryside Commission is an autonomous Government agency. However, the greater powers and resources available to the NCC enable it to go beyond the advisory role of the Countryside Commission and purchase and lease sites of wildlife value.

The work of both the NCC and the Countryside Commission is assisted and supplemented by the activities of local authorities and voluntary bodies. Both organisations provide grants to local authorities for various schemes such as Country Park and Local Nature Reserve (LNR) management in which they have an interest.

The National Trust plays an important role in safeguarding scenic areas and this complements much of the Commission's work. Various organisations, particularly the Royal Society for the Protection of Birds (RSPB) and local naturalists' trusts work closely with and are aided by the NCC.

3. Coastal organisations

3.1. Heritage Coasts

Management objectives: Heritage Coasts administered by the Countryside Commission and local county councils, essentially have two-fold objectives:
   i) to conserve the quality of scenery,
   ii) to foster leisure activities which rely on natural resources.

In respect of conserving environment resources the aim is 'to make the wisest use of all coastal resources rather than to preserve scenic stretches for their own sake or to discourage access thereto'. (Countryside Commission, 1970 p. 16.)

Cullen (1982) identified a number of other unstated aims of the scheme centered around the pioneering nature of the Heritage Coast concept (they were founded in 1974) through which the Countryside Commission could gain experience relevant to countryside conservation management generally. In particular it was a testing ground for such management tools as project officers' management plans, grant-aiding and voluntary agreements and provides a stimulus for 'communication and co-operation between local authorities and other bodies responsible for managing particular ecosystems of interest' (Cullen, 1982). Fostering of good will between such parties could be hoped to encourage positive management initiatives for other areas. Two such coasts occur in the area: Glamorgan which was founded in 1974 and Gower, founded in 1982.

Guidelines for managing Heritage Coasts are set out in 'The Coastal Heritage' (Countryside Commission, 1970). The document established the following basic management principles (Williams, 1987):
(a) Determination of intensity of use.
(b) Management zones based on different intensities.
(c) Control of development.
(d) Regulation of access.
(e) Landcape improvements.
(f) Diversification of activities.
(g) Provision of interpretative services.

As there is no provision for the statutory designation of Heritage Coasts the Countryside Commission has an important role to play in persuading local authorities to define Heritage Coasts within structure and development plans and

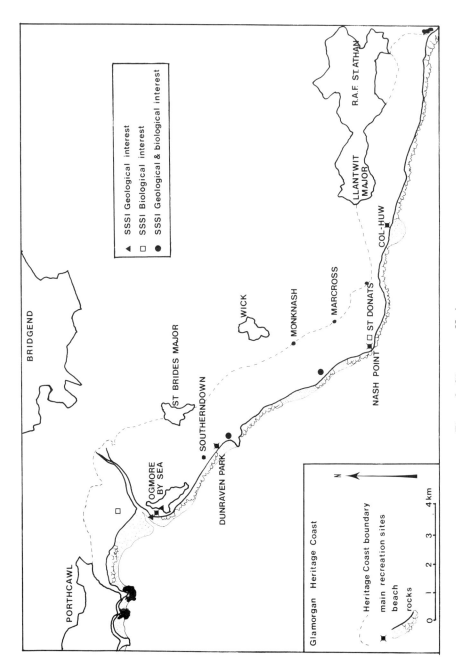

Figure 3. Glamorgan Heritage coast.

establish management planning. The Commission can provide advice on all matters relating to definition and management and gives financial incentive.

Functioning of Heritage Coasts: Local authorities are expected to appoint an Heritage Coast Officer and provide him/her with such professional back-up from their various departments as required. The Officer should have at his/her disposal a small fund to implement small scale improvements and Councils take steps to facilitate the implementations of measures contained within the management plan.

The Countryside Commission places great emphasis on the need to prepare management plans. These documents identify recreation patterns along the coast and outline a zonation policy that ensures the non-distribution of sensitive ecological sites whilst easing visitor pressure in congested areas. The task of drawing up these plans usually falls to the Heritage Coast Officer with assistance from appropriate council departments.

Provisions in the 1968 Town and Country Planning Act eventually enables proposed Heritage Coast plans to become incorporated into district and structure plans. The district plan indicates policy towards each of the management goals and objectives as well as the precise boundaries, both lateral and inland, once these have been established. Designation proposals are subject to the approval of the Secretary of State for Wales in the case of Welsh Heritage Coasts.

Project officers are a key element of the Heritage Coast concept. Their duties are primarily to draw up a management plan, organise practical improvement works along the coast and negotiate with local farmers, landowners and interest groups with a view to securing support for the programme and establishing voluntary agreements for such matters as access and car-parking. Such agreements are an essential part of policy implementation. But where they cannot be secured or fail to work there is provision for compulsory purchase to bring key areas into public ownership, e.g. Col Huw beach (Figure 3). Liaison with the National Trust, landowners and farmers should seek to establish codes of practice for agricultural and other operations (Cullen, 1982). Furthermore Project Officers are expected to oversee the preparation of interpretative and promotional material, to recruit wardens and clerical staff as required and to instigate a forum for community representation. They are also required to submit quarterly reports to the Commission and to assume general responsibility for the planning and day-to-day running of the scheme. In all actions the Officer is responsible to the advisory management committee.

Supplementing funding provided by the Commission and local authorities has proved to be an important function of the Officer (Cullen, 1982; Howden and Williams, 1985). A wide range of grants is available from Government agencies, particularly Tourist Boards and the Manpower Services Commission (MSC). The latter is able to provide support for Youth Training Schemes and Community Programmes to carry out improvement (but not maintenance) works along the coast and to support particular posts for a period of not more than one year, although the scheme can be renewed every year. Furthermore, through developing community support a variety of fund-raising activities can be organised and local businesses may be persuaded to provide free services. Community support is not just required to boost the available budget. The fostering of good will is vital and the concept recognises the benefits that accrue by developing support from both residents and visitors alike.

Once programme implementation has commenced, wardens are employed to patrol the coast, giving information and assistance to the public and enforcing byelaws, particularly those relating to the Wildlife and Countryside Act. They also act as information gatherers and carry out maintenance and improvement tasks. In all these activities they are assisted to varying degrees by volunteers who are either local individuals or come through various institutions, especially schools, colleges and youth groups.

3.2. The role of the Nature Conservancy Council (NCC)

The present guise of the NCC as an autonomous national government agency was established by the Nature Conservancy Act, 1973. Under this Act the functions conferred upon the NCC were

i) to establish, maintain and manage National Nature Reserves;
ii) to advise ministers on policies for or affecting nature conservation in Great Britain;
iii) to provide advice and disseminate knowledge about nature conservation;
iv) to commission and support or if necessary carry out relevant research;
v) various duties under other statutory provisions particularly the notification and protection of SSSIs (NCC, 1985, p. 68).

NCC policy. Moore (1982) described the aim of nature conservation as safeguarding the national heritage for the enjoyment of present and future generations. This is achieved through promoting an awareness for wildlife and its special needs and by designating protected sites whereby native plants and animals may continue to perpetuate within their viable habitat range. National Nature Reserves (NNRs) and Sites of Special Scientific Interest (SSSIs) have become the instruments of protection. The distinction between them is a matter of control. The NCC is directly involved in the management of NNRs and will either purchase the freehold or leasehold of the land or will enter into a formal management agreement with the owners and occupiers. To date outright purchase of land has afforded the best protection for NNRs (NCC, 1984).

The designation of SSSI was used either as an interim protective measure prior to the establishment of NNRs or as a means of providing protection to sites harbouring rare species or having some other noteworthy feature, including geological significance. Originally SSSIs were merely notified to planning authorities who had then to consult with the Conservancy where development proposals would have an impact on their interest; where conflicting interests arose these were usually brought before local planning inquiries for adjudication. But under the terms of the Countryside Act (1968) the NCC was empowered to enter into management agreements with the owners of SSSIs although lack of funds and planning control over agricultural and forestry practices limited their effectiveness.

The Wildlife and Countryside Act (1981), despite loopholes which left many proposed SSSIs vulnerable to damage prior to notification, reinforced the status of the designation by obliging the NCC to re-notify all owners and occupiers of SSSIs with a complete list of activities detrimental to their interest value. Where curtailment or cessation of notified activities incurs a loss of revenue to the owner or occupier, the NCC is able to award compensation payments. However the co-operation of owners is still essentially on a voluntary basis and SSSI destruction and damage continues at an alarming rate. Between April 1984 and March 1985, 255 U.K. sites were damaged, 94 of which will in whole or part require denotification. In over half these cases the damage was a result of agricultural activity (NCC, 1985).

NNR and SSSI selection. The complicated process by which both NNRs and SSSIs have come to be selected reflects the complexity of natural communities. Consideration needs to be given not only to site quality but also to the frequency with which similar sites occur within a given region. For this reason a surviving example of an ancient mixed woodland in England may merit NNR status whereas a biologically richer site in Gwent where such woodlands are relatively common may not. Strategic designation of this nature aims to conserve the geographical distribution of species and habitat types. The size of designated areas is obviously of great importance for many species require large areas if they are to retain viable populations. Big is better is an important maxim for conservation and most SSSIs will represent the largest and finest surviving examples of their kind in each region. Conditions vary with species, as does the knowledge required to make such decisions but a typical guideline would be to select SSSIs from within geographical areas of between 60 and 400 thousand Ha with a spacing interval for particular habitats of about 50 km (Moore, 1982). These figures are adjusted in accordance with new data from ongoing research programmes. Site designation is a dynamic process; new sites are constantly being added and old sites denotified when through human activity or natural processes they no longer merit protection. Other sites may be selected where they are threatened by development or represent a rare or a unique habitat or provide sanctuary for endangered species. Gower alone has 24 SSSIs, 10 being geological. Their size ranges from 0.5 to 6000 Ha.

Apart from its responsibilities for the notification of SSSIs and conducting research to this end the NCC also provides grants to appropriate Non-Government Organisations (NGOs) and assists local authorities in establishing Local Nature Reserves (LNRs). Grants payable to NGOs vary according to need. For the financial year 1984–85 for instance the Glamorgan Trust for Nature Conservation received a grant of £4421. The total award made to U.K. NGOs for the same year was in excess of £330 000 (NCC, 1985).

The function of Local Nature Reserves. Section 21 of the 1949 Act allowed local authorities to establish LNRs after consultation with the NCC. Their status is equivalent to that of NNRs and they also receive protection as SSSIs, but their context is local and amenity use tends to assume a higher priority in their management.

One hundred and five LNRs had been established in the UK by 1983 (NCC, 1984). They are largely financed by local authorities, grant-aided by the NCC. Kenfig Dunes and Pool LNR (Figure 2) for instance received grant support towards salaries in its formative years from the NCC which also provided £5000 towards the building of an information centre and other occasional grants. Gower has 18 LNRs within its boundaries.

i) Oxwich National Nature Reserve (Figure 1). In the past, site preservation and the use of NNRs as outdoor laboratories have been the overriding concerns of the NCC with respect to the management of NNRs. Recently increasing emphasis has been placed on the need to cater for visitors. Oxwich has become the foremost NNR within the South Wales Region and is now in the process of maximising the use which visitors can make of the reserve within the constraint of safeguarding site diversity.

The site covers 750 acres and incorporates a wide range of habitats from freshwater marsh to sand-dune and woodland. Most of the site has free access with the notable exception of the freshwater marsh which is greatly restricted and the enclosure of the less stable dune areas.

Most of the reserve is owned by the NCC but parts of it are held on a lease arrangement or formal nature reserve agreement. The car park area is beyond NCC control and revenue from car parking charges is not therefore ploughed back into the reserve. A consequence of this is that the car park area is an obtrusive eyesore which is unfortunate in an otherwise beautiful bay.

The reserve is staffed by a warden and an assistant, helped by a four-man MSC scheme to carry out 'maintenance' work. The reserve centre is utilised as a classroom for 10 months of the year under the Gower Field Education Project and a teacher working from the centre is employed by West Glamorgan County Council for this purpose. A clerical officer is also employed.

Facilities laid on for visitors to the reserve are limited. Various leaflets and guides have been prepared covering introductory and more advanced features of the site and a series of excellent information boards has been erected outside the centre. There are also two nature trails, though they may be scrapped in favour of less formal exploratory devices.

Many visitors to the site come only for foreshore recreation, unaware or heedless of the functioning of the NNR. There is no attempt to discourage such use though wardening of the dune enclosures is necessary to stop trespass. For this category of user the information and educational services provided at the Centre would seem to lack the stimulus needed to encourage them to become interested in the ecological value of the reserve. In the past, foreshore attraction has kept the majority of visitors away from the reserve. This has been in the best interest of the site but at the same time suggests that encouraging greater use of the reserve would not result in its being inundated by visitors. Although access to most parts of the reserve is unrestricted, the problem remains of whether the identified need for greater visitor facilities can be reconciled to the maintenance of site interest and diversity. It is a problem which can only be resolved by a site-specific approach with careful monitoring of visitor patterns and site degradation over a phased introduction of improvements. It is this need to which management is now addressing itself.

The project has proved to be a great success and has flourished under a Local Authority committed to countryside conservation and conservation bodies with an enlightened approach to visitor management. By 1986 35 000 schoolchildren had made use of the centre and over 40 schools and colleges now make regular use of the service. Much work remains to be done in the development of hides and other facilities, but despite cutbacks in resources commitment to the project is unswerving.

ii) Kenfig Dunes and Pool Local Nature Reserve (Figure 1). Between Port Talbot Steelworks and Porthcawl lies a sand dune system of moderate extent, some 1500 acres. The western area of dunes is known as Margam Burrows and is dominated by the steelworks, while the eastern end known as Kenfig Burrows is a remote dune system which incorporates a freshwater pool. It is the Kenfig Burrows which afford the greatest wildlife interest and a large part of them, right down to the shore, has been designated as a Local Nature Reserve. The reserve, established in 1978, is comprehensively managed by Mid Glamorgan County Council which employs a project officer responsible for its day to day management. Total funding from the council amounts to some £30 000 per annum.

The lease for the reserve is held by the council and the project is run by a management committee comprised of officers of the council's planning department. The project officer is assisted by a warden who is supported on an MSC scheme. A management plan recently prepared (1986) is awaiting council approval.

Education is seen as an overriding priority of the scheme, teaching local people and visitors how to enjoy and look after the area.

A number of bodies have interests in the area, especially the NCC which provided grants for salaries in the early years of the scheme and contributed £5000 towards the centre.

Whatever the title Local Nature Reserve may suggest, the site still fulfils an important educational role. Kenfig Pool, for instance, is used both for fishing and canoeing, while a local gun club has the shooting rights for the dunes (see section 3.6). This latter activity particularly would appear to be hard to reconcile with the conservation of the site but the club respects the views of the warden and its members behave accordingly. The success of the management approach in this respect shows that the principles of conservation can still be met whilst enabling the site to accommodate a number of potentially conflicting recreational pursuits. It is likely that without management presence to curb the excesses of all users and minimise disturbance to sensitive areas the site would rapidly deteriorate. Although recreation on the site is not promoted its importance for this purpose is clearly recognised and as such is planned for.

Community involvement in the programme has been developed through a Watch Club for youngsters and by arranging talks and guided walks. Volunteers have been encouraged to take part in the scheme and the centre is now able to open on weekends, when visitor use is highest, as a result of volunteers providing necessary staffing.

Links with other schemes in the region have been developed on an informal, ad hoc basis as a result of contact between wardens of the various schemes. There seems little attempt to develop such contacts at higher levels particularly across county boundaries.

Since its inception in 1978 the scheme has demonstrated how the conservation interest can be safeguarded. Simple parameters such as the successful breeding of great crested grebes and mute swans since the introduction of the wardening service are as useful indicators as any of its cost effectiveness. On a more practical level there is now less flower-picking; motor-bikes and horses are effectively controlled and limited improvements to the car-parking area have increased the capacity to 420 cars.

As the scheme becomes better known visitor usage increases and although attempts have been made to attract visitors to the site this has been done in a low-key fashion in a way that develops local interest and understanding of the significance of the site and the need to conserve it.

Coastal erosion, though recognised, is beyond the financial means of the scheme to counteract and is manifested in the disappearance of finer beach material over the years. Elsewhere dune erosion is not a great problem and visitor trampling is often of benefit to many species of flora, particularly orchids.

Recreation and the NCC. Early in its history the NCC established credentials as a body committed to wildlife conservation based on a firm foundation of scientific rigour. The acquisition of NNRs enabled it to undertake large-scale outdoor research programmes without interference. This continues to be an important use of NNRs but increased recreational pressure has forced the NCC to devote more energy to education and interpretative services. The enclosure of wild habitats from human activity cannot of itself form a justifiable land use except over strictly limited areas. Most conservation sites have to accommodate if not provide for amenity use. Careful management can usually prevent undue damage to sites arising from high levels of use and well organised educational services can both enhance amenity value and foster environmental awareness. This has been shown to work well in Gower, for instance, where West Glamorgan County Council and the NCC have worked together to develop the successful Gower Field Education Projection. Yet despite acknowledging the importance of recreation – 'nature is now a recreational and tourist asset of the greatest importance' (NCC 1984, p. 39) – the NCC has failed to provide effective guidelines by which intensive visitor use can be reconciled to site conservation.

Coastal sites. Although there are many coastal NNRs and SSSIs, 650 of which in the UK include portions of intertidal zone, until 1981 there was no provision for statutory protection of sublittoral communities. Under the 1981 Act however, provision was made for the establishment of Marine Nature Reserves (MNRs). Attempts by the NCC to establish the first of these have been beset by difficulties and so far little has been achieved beyond the proposal of seven possible sites, one of which is Lundy Island, in the Severn Estuary. Conservation of the wider marine environment remains in the province of National Government and particularly the MAFF.

3.3. Coastal management and ownership objectives of the National Trust

The National Trust has a considerable land holding (5000 acres) in the Gower Peninsula and like the NCC is therefore prominent in its management. A warden is now employed to manage the Gower property and is presently engaged in formulating a management plan for the area in which landscape conservation will assume a priority. While the Trust is exclusively responsible for managing its own property there is considerable co-operation between it and the various other conservation bodies also at work within the area. The need for such close links is fully accepted if only because in many instances the Trust does not control (those all-important) points of access, thereby making liaison with other bodies essential.

The Countryside Commission provides 50% of the cost of funding for the warden and similarly meets 50% of the cost of many management projects. The Commission also provides assistance with major land acquisitions. Practical improvement works are undertaken with MSC labour (£30 000 per annum). Future developments include the preparation of site-specific management plans and establishing the role of Rhosili Visitor Centre as a major interpretative base rather than its previous function as a gift shop (Figure 2). This reflects the growing concern the Trust has with countryside management and the need to promote visitor awareness.

Stretches of U.K. coastline – they now own 481 miles of U.K. coast acquired at an average cost of £15/foot – were amongst the first properties acquired by the National Trust. Since then the Trust's coastal holdings have increased considerably; 75 miles of coastline had been acquired by 1965 and this was dramatically boosted to almost 450 miles by 1985 following the launch of Enterprise Neptune in 1965. A re-launch of Enterprise Neptune in 1985 is expected to extend considerably the Trust's coastal acquisition and management programme. The Gower area has the largest collection of Enterprise Neptune properties in England and Wales.

Because of the inalienable nature of Trust property such a recent massive drive to protect scenic coasts should be welcomed and yet much criticism has been voiced over the Trust's failure to provide management initiatives for its land holdings, particularly along the coast. Cullen (1982) for instance hinted that 'the management structure the Trust uses for its land holding should be strengthened to help cope with the considerable expansion to their land holdings' (Cullen, 1982, p. 69). Cullen (1982) also comments on the paucity of interpretative material at coastal sites and the failure to bring to public awareness a strategy for coastal management. Since 1985 however, the Trust has assumed a more positive attitude towards management and now expects to draw up site-specific management plans for all their coastal holdings and to provide appropriate wardening and interpretative services usually with advice and financial assistance from the Countryside Commission.

Burgeon and Hearn (1984), both senior Trust officers, noted that 'there is an increasing emphasis within the Trust on interpretation of its coast and countryside properties, informing the public on management in progress ...' and that 'the subject of interpretation on each coastal property, along with all other aspects of the Trust's management and responsibility will be carefully considered during a major three-year programme of Management Plan preparation, due to start in the autumn of 1984' (Burgon and Hearn, 1984, p. 24).

Already there are signs that the Trust is acting on its words. A recent paper (National Trust, 1985) listed the provision of management plans for all their landholdings as the major priority for the use of resources and this is being taken up at ground level at more and more sites.

While acquisition of fresh holdings in U.K. is continuing at an average rate of one per week, increasingly the need for definitive management of each site is being appreciated. Establishing and maintaining management programmes is very costly and careful consideration needs to be given to the priority attached to new acquisitions as opposed to providing further management services.

The criteria used for selecting sites suitable for acquisition to be held inalienably are that the area in question must be of outstanding scenic value or of historical and archeological interest and must be of national importance. Where the land is of less than outstanding scenic value the decision to acquire may be swayed by its historical, archeological or wildlife significance. However, the Trust recognises the roles of other conservation bodies and realises that such organisations as the RSPB, NCC and Naturalists' Trusts may be better custodians of many sites, and the need for close co-operation between the various amenity and conservation bodies is accepted.

Properties are generally managed through the Trust's regional offices, although some are managed by local committees, and a smaller number by Local Authorities. Management is now seen as most effective when under the control of the regional offices, e.g. Llandeilo in the context of South Wales.

Management plans are correctly seen as essential tools, identifying management objectives, providing continuity, noting valuable areas and establishing a framework for the efficient use of resources. They should outline the work to be carried out and the resources required. As such they are directly comparable to those envisaged for Heritage Coasts. With 50% grant-aiding from the Country-side Commission, all the Trust's regions are currently engaged in the production of management plans. The target date for their completion is the end of 1987. The Land Management Agent is responsible for writing the plans, assisted by a Management Team and approved by the Regional Committee. Plans are expected to be short and should be reviewed every five years.

Executing the plans is the responsibility of wardens who are responsible to the Managing Agent. Large properties employ a Head Warden full-time and seasonal wardens at his command. Wardens' roles are little different from those employed under Heritage Coast schemes. Improvement works may be carried out by the wardens themselves or by MSC schemes or volunteer and contractual labour.

Trust policy in areas such as AONBs, Heritage Coasts and National Parks is to co-operate and where possible establish small working parties or advisory groups.

Points of access are seen as critical in management programmes and policy here is to control these by direct ownership wherever possible. In addition to the control of visitor levels that this affords, ownership of access points also provides opportunities to gain revenue from visitors and to act as recruitment grounds.

Much to their credit the Trust has established detailed monitoring and survey schemes on many of their sites. These have provided information on how to minimise visitor impact in a manner which has brought positive benefits to visitors. Although nature conservation is not seen as an overriding concern of the Trust the old Laissez-faire attitude is now being replaced by a more positive management approach to the welfare of wildlife. Obviously this drive towards establishing adequate management programmes on all its landholdings is a costly enterprise for the Trust to undertake. Finance is derived from four sources: grant-aid, appeal fund, subscriptions and income from visitors. A number of government agencies provide grant-aid, of which the Countryside Commission is the most important. To date the Commission has provided in excess of £750 000 towards acquisitions and provides 50% grants for wardening and practical works. Since 1981 the NCC has provided increasingly large funds for nature conservation work.

3.4. County and district council responsibilities for conservation and the coastline

Planning duties are divided between county and district councils. A devolutionary process over recent years has provided district councils with a more authoritative role in countryside issues than previously. Overall responsibility for rural protection remains with county councils through the development of Structure Plans, but within Local Plans district councils provide the detailed in-filling of policy by which Structure Plan objectives are met.

Policy statements prepared jointly or individually by councils relate to specific localities or topics and supplement structure and or local Plans. The Glamorgan Heritage Coast Plan statement is one such document, prepared jointly by two borough and two county councils (Williams and Howden, 1979).

As the statutory coastal protection authorities, district councils have a particular interest in coastal planning and are therefore likely to play an important role in coastal conservation schemes. The uptake of management programmes in Heritage Coasts, AONBs, County Parks and LNRs depends largely upon council initiatives. Where councils are sympathetic towards conservation, e.g. South Glamorgan County Council, and can allocate sufficient resources for this purpose then the aims of such designations can be effectively met.

In addition to management programmes within recognised designated areas, councils may establish conservation programmes in other areas where recreational use is already well established. Such initiatives, e.g. Caerphilly Mountain Project (Figure 1) which arose directly from the observed successes of the Glamorgan Heritage Coast programme, may also qualify for Grant-aid from the Countryside Commission or NCC. Moreover certain areas might be singled out for special protective measures as for instance in the Barry-Penarth local plan where the coastline between the two towns is defined as a coastal conservation area.

The two Country Parks considered here, Margam and Cosmeston Lakes, differ substantially to the extent that the management of Margam Country Park is wholly incompatible with the ethos of Heritage Coasts, for instance, while Cosmeston although established for recreational use is a more typical countryside management programme. That both these schemes are supported by the Countryside Commission is not surprising as a recent statement notes, 'The Commission has promoted the assistance it can offer but has not done so in a rigid programmed fashion. Assistance has therefore been sensitive to local perceptions of need and opportunity. In consequence the pattern of provision of country parks ... varies quite considerably from region to region and from county to county'. (Countryside Commission, 1986, p. 7).

Cosmeston Lakes Country Park was established on the site of old quarry workings in 1975. Cosmeston Lakes Country Park has been successful both as an exercise in land reclamation and as an attempt to provide recreational facilities catering for the large nearby population of Cardiff and its environs. The Park is funded jointly by the Vale of Glamorgan Borough Council and South Glamorgan County Council and receives grant support from the Countryside Commission. The Vale of Glamorgan Borough Council is responsible for its administration and the appointment of wardens. Currently the staff consists of a warden and assistant warden. An MSC scheme employing about 14 youngsters is currently engaged on an archeological dig and the reconstruction of a medieval village in one corner of the site.

The Park contains two lakes, one set aside as a wildlife sanctuary and the other used for boating and swimming, and it is this that provides the main attraction with most visitors tending to congregate around the shores. Nonetheless the remainder of the site is of considerable wildlife interest and it is likely that selected areas may shortly be notified as SSSIs.

Despite the relatively straightforward administrative set-up, management has suffered as a result of antagonisms between the two council departments responsible. This has resulted in the loss of a £35 000 Countryside Commission grant towards the improvement of toilet and office facilities at the site through a failure of the two councils to commit sufficient funds of their own to the cost.

In accordance with Countryside Commission stipulations a management plan for Cosmeston has been drawn up by South Glamorgan County Council with assistance from the Vale of Glamorgan Borough Council. Now that reclamation of the land is complete and the Park boundaries have been fully established with access to all areas, the proposals within the plan have been largely executed. Community and volunteer involvement is limited. School groups do use the site for wildlife project work.

3.5. Glamorgan Trust for Nature Conservation

The Glamorgan Trust for Nature Conservation controls over 40 reserves with a total land area of about 1300 acres spread throughout the counties of West, Mid and South Glamorgan. While many of these are inland they do look after a number of coastal sites, particularly in the Gower, and are further involved in coastal conservation through their active support of other schemes such as the Kenfig Dunes and Pool Local Nature Reserve and indeed the Glamorgan Heritage Coast. They also co-operate closely with the NCC in many of its wildlife projects, such as bat surveys, and its members may act as wardens for many small SSSIs which the NCC could not reasonably expect to cover without a large increase of staff levels.

Reserves may either be owned outright or leased on a long or short term basis. A number are recognised SSSIs and these receive support from the NCC towards their maintenance. Every site has a volunteer warden appointed to it and in over 50% there is provision for public access.

Although the stated objective of the Trust is 'to develop and improve the capability for practical wildlife conservation and protection' (Glamorgan Trust for Nature Conservation, 1986, p. 1), the Trust recognises the need to provide an educational/interpretative service for visitors and to this end is planning to appoint an educational officer and establish an interpretative centre in the Gower.

The service which the Trust provides for the local community is essentially the conservation of small areas of land whose wildlife value though significant does not merit their protection as SSSIs. Indeed there is now a general policy of increasingly extending tenureship to such areas rather than those of greater national significance.

Grant support comes from a number of sources. The NCC for instance provided over £4000 towards specific projects in 1984/85. Local Authorities also provide a source of income although amounts vary from council to council. Grant support from Mid Glamorgan County Council tends to be less good. Support from West and South Glamorgan Cuonty Councils tends to be readily forthcoming (Glamorgan Trust for Nature Conservation, 1986). Councils may also give support in kind, by making favourable terms available for sites on council land such as Tremains Wood.

Regionally the Trust fulfils something of a watchdog role for conservation interests and is able to take an independent line where necessary. For the recent (1986) Merthyr Mawr Planning Inquiry, for example, the Trust provided its own submission against the proposals.

Sites are surveyed and managed according to their perceived needs. There is no overall strategy for acquiring sites, acquisitions being made as and when land becomes available on suitable terms. Similarly there is no specific coastal policy although great importance is attached to Trust-held land in the Gower where an interpretative centre is planned.

3.6. Non-Government Organisations (NGOs)

By far the most important non-government organisation involved in conservation of coastal scenery is the National Trust whose role has already been discussed. But there are a number of NGOs involved in wildlife conservation along the coast, these including the Royal Society for the Protection of Birds (RSPB) and the Royal Society for Nature Conservation (RSNC) which is the umbrella organisation for county naturalists trusts. Both these organisations manage coastal reserves with the aim of preserving their wildlife interest or more particularly in the case of the RSPB their ornithological interests. Such reserves are usually leased or owned and are run and maintained largely by volunteers. Funds are raised mostly by subscription, supplemented by NCC and local authority grants, and private donations and bequests.

Royal Society for Nature Conservation. The RSNC is essentially the national co-ordinating body for Nature Conservation Trusts, such as indicated in section 3.5 above. Acquisition and management of reserves is a matter for individual Nature Trusts. Forty-six Trusts have now been established covering the whole of the U.K., with a total membership of 150 000. The Glamorgan Trust for Nature Conservation (GTNC) was formed in 1961 (as the Glamorgan Naturalists' Trust) by a group of concerned people to conserve the wildlife in the three counties of Glamorgan. It is a self-financed charity dependent on membership fees and voluntary contributions. A few of the reserves are in coastal areas, such as five sites around the coastline of the Gower Peninsula and one site at Lavernock Point. The GTNC is still expanding and will acquire new sites in the future. The main Trust aim is conservation of wildlife. Because of this, the sites are not advertised although open days and guided walks act to show people practical examples of conservation in action and the importance of each reserve as a refuge for wildlife. Although a few positions such as Warden are full-time paid jobs, many positions are voluntary or part of Manpower Services Commission Schemes. Members of the Trust are invited to help in reserve maintenance as well as talks, excursions and field meetings. Reserve maintenance is also carried out voluntarily by schools, youth organisations and the British Trust for Conservation Volunteers (BTCV).

The Trust, as well as managing their reserve, acts as an educational and advisory body. The Glamorgan Nature Centre, the headquarters of the Trust, was built at Tonall, near Bridgend, and opened in 1982 by Prince Charles. The Centre is surrounded by landscaped ground, the Park Pond Nature Reserve and Cwm Risca Wood Nature Reserve on the nearby Countryside Commission Demonstration Farm. The Centre runs a WATCH group for young people aged 8–18 to educate them to the need for environmental management. Advice from the Trust can be gained on wildlife conservation and management both in countryside and urban areas. It also carries out surveys and takes part in national surveys such as the successful 'Common Butterfly Survey' started in 1984.

Kenfig and District Wildfowlers' Association. The Kenfig and District Wildfowlers' Association (KDWA) was formed in 1965 and is affiliated to the British Association for Shooting and Conservation. The group has a membership of 50, and a waiting-list for prospective members. The membership fee is £25 per year, so making it self-financed, plus members have to pay £6 a year for a wildfowl licence. A licence allowing the Wildfowlers to shoot within the Kenfig Pool and Dunes LNR is granted by the Kenfig Trustees every five years.

Although the membership to the Association is 50, a maximum of 25 is allowed at any one shoot. When the group was established they shot six days a week, but since the Nature Reserve has been established this has been reduced to two. Throughout the shooting season the KDWA pursue their sport every Wednesday and Thursday. The area in which shooting is allowed has also been reduced to 315 hectares since the Kenfig Nature Reserve was established. These limitations and reductions have mainly been through voluntary agreement, and in some cases were self-imposed, but combined they have led to a general feeling of unfair treatment by the Nature Reserve. The KDWA felt that as the area was rural common land and they have been together far longer than the Nature Reserve has been present, they have been unfairly pushed out. The Wildfowlers believe that when they shot 'vermin' such as magpies, foxes and rabbits on a more frequent scale, the numbers of plovers and songbirds in the area became high. Since the reduction in shooting days was agreed the numbers of these birds has decreased. To prevent depletion of 'vermin' the Kenfig Wildfowlers do annually introduce ducks and pheasants into the area, and they maintain that they actually shoot far less than the stocking level. They also run a clay pigeon facility.

There is obviously an important question of safety where guns and live ammunition are concerned. Although there have not been any injuries, the safety of the general public is of paramount importance to the Kenfig Wildfowlers and every effort is made to maintain their untarnished safety record.

4. Conclusions

The South Wales scenario epitomises what is happening at many U.K. coastal areas. The coastal strip is a dynamic entity and several agencies have inputs into its management structure. Maintaining coastal quality is difficult as various agencies tend to internalise problems which are effectively inter-agency problems. Issues marginal to the core agency need tend to be ignored until crisis levels are approached. This is singularly apt when financial constraints are necessary, i.e. the financial climate of the mid-1980s. Boundaries are drawn delineating various coastal areas but in practice natural processes in this area are influenced by homeostasis. The cultural (human) dimension is dominated in the main by Markovian processes whereby current decisions reinforce earlier ones. At local levels, much informal information exchange exists, but on the loftier national plane replication is frequently common.

Too much emphasis is placed on the volunteer sector and the Manpower Services Commission. Current (1987) thinking is that recruitment for coastal management schemes under MSC control will be much more difficult, will be carried out by an older workforce which has been unemployed for 12 months; and more rigorous training will be introduced into the schemes. This has enormous implications for amongst other things the maintenance question. It is highly unlikely that a new agency will overview the coastal zone but a national strategy should be adopted, focusing on the coastal zone, if only to itemise priorities. A condition of multiple use proliferation and isolated perspectives currently exists rather than priorisation. This should include the area seawards of the land, an area not considered in this paper.

Acknowledgements

I would like to express my deepest thanks to the many midnight discussions held with the late Professor O. A. C. Williams which provided me with many ideas relating to U.K. coastal management.

References

Burgeon, J. P. and Hearn, K. P., 1984, National Trust Heritage Coasts. RERG Conf. Proc., 24.
Countryside Commission 1970, The Coastal Heritage. London: HMSO.
Countryside Commission 1986, Recreation 2000: A discussion paper on future recreation policies. Cheltenham: Countryside Commission, 7.
Cullen, P. 1982, An Evaluation of the Heritage Coast. Programme in England and Wales. Cheltenham: Countryside Commission, 46, 69.
Glamorgan Trust for Nature Conservation 1986, 1986–89. 3 Year Plan. Bridgend: The Trust, 1.
Howden, J. and Williams, A. T., 1985, A new approach to Coastal Management: The Glamorgan Heritage Coast Experience. *Proc. Coastal Zone* **85**, 26–35.
McEwen, N. and A. 1982, National Parks, Conservation or Cosmetics? London: George Allen and Unwin Ltd.
Moore, N. 1982, What parts of Britain's countryside must be conserved? *New Scientist, Jan 21*, 147–149.
National Trust 1985, Coast and Open Country: First Review. Cirencester: National Trust.
Nature Conservancy Council 1984, Nature Conservation in Great Britain. Peterborough: NCC, 68.
Vale of Glamorgan Borough Council 1980, Barry-Penarth Coastal Plan: Report of Survey – Summary. Barry: The Council.
Williams, A. T., 1987, Coastal conservation policy development in England and Wales, with special reference to the Heritage Coast concept. *Journal of Coastal Research* **3** (1), 99–106.
Williams, A. T. and Howden, J. C. 1979, The search for a coastal ethos: A case study of one of Great Britain's 'Heritage Coastlines'. *Shore and Beach*, July 1979, 17–21.

Allan T. Williams
Coastal Research Unit,
Department of Science,
The Polytechnic of Wales,
Pontypridd,
Wales,
U.K.

3. Recreational uses and problems of Port Phillip Bay, Australia

Introduction

Port Phillip Bay (Figure 1) in south-eastern Australia has an area of about 1920 square km, a maximum depth of 24 m, and a coastline 256 km long. It is a marine embayment that receives rivers and creeks from a hinterland catchment of 9600 square km, the largest river being the Yarra, which enters the head of the bay. Mean tide range is about 1 m, and waves are generated by winds mainly from the north, west and south. Occasional south-westerly gales produce short, steep waves up to 3 m high on the bay, and these break heavily upon the eastern coastline. Winds and tides generate moderate currents within the bay, but there are strong tidal currents form 'the rip' as tides enter and leave through Port Phillip Heads.

The city of Melbourne (population 2.9 million in 1986) has developed on the northern shores of the bay, and its suburbs extend along the coastline south-west to Altona, and south-east to Frankston, with holiday resorts and residential settlements continuing intermittently round the east coast as far as Portsea. The city of Geelong (population 180 000 in 1986) stands on the south-western coast, at the head of Corio Bay, and there are small seaside resorts on the Bellarine Peninsula, to the south. Melbourne and Geelong are major ports, servicing more than 3000 commercial ships which enter and leave Port Phillip Bay each year, and there are substantial industrial areas, especially to the north of Geelong, and from the port of Melbourne on the Yarra estuary south-west to Altona: about 70% of the factories established in the state of Victoria are located in the Melbourne area.

The coastline of Port Phillip Bay is varied (Figure 1). Much of the east coast is steep and cliffed, with rocky shores and beaches of sand and gravel mainly in coves and embayments, but there are low-lying beach-fringed sandy plains on the northern shore, between the Yarra mouth and St Kilda, on the eastern shore between Mordialloc and Frankston, and on the southern shore between Safety Beach and Sorrento. The west coast between Geelong and Melbourne is low-lying, a basaltic plain with areas of salt marsh and alluvium, with shelly beaches along much of the shore, but the coast of the Bellarine Peninsula is generally higher and steeper, with narrow beaches of sand and gravel. Spits of sand and gravel border Swan Bay, a shallow embayment bordered by wide salt marshes, and there are sandy beaches, interrupted by headlands of dune calcarenite, on either side of Port Phillip Heads, the marine entrance from Bass Strait.

The floor of the Bay declines from the eastern and western coasts into a broad, almost flat-floored basin up to 24 m deep, but the southern part is a shallow sandy threshold, with extensive shoals exposed at low tide. Mud Islands, the only outlying natural islands, are dune-capped sand barriers that enclose a lagoon and salt marshes. Between the shoals, deeper channels maintained by tidal scour converge

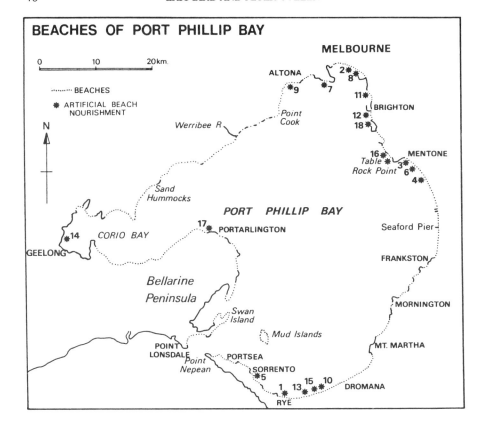

Figure 1. Map showing the extent of recreational beaches on the coastline of Port Phillip Bay, Australia, with the location of the 18 artificially nourished beaches.

southwards to Port Phillip Heads. The combined channel winds and deepens to more than 70 m: it has been made less dangerous for navigation by the blasting of submarine pinnacles of dune calcarenite, but there are still strong currents as the tides rise and fall, and when these interact with strong wave action from Bass Strait navigation through Port Phillip Heads becomes hazardous, especially for small boats.

There have been minor changes on the coastline bordering Port Phillip Bay during recent decades (Bird, 1974). Some cliffed sectors have receded, and in places sea walls and boulder ramparts have been introduced to halt this recession, especially where it threatened to undermine developed property or the coastal highway. Beach erosion has become a problem around Port Phillip Bay, many of the beaches having been reduced in extent over the past few decades. Beach erosion is a widespread phenomenon in Australia and overseas (Bird, 1985): here it is mainly due to the building of sea walls to halt cliff recession (thereby cutting off natural sand replenishment from eroding cliffs) and breakwaters that have inter

Figure 2. Artificial beach emplaced in 1984 at Black Rock, Port Phillip Bay, provides opportunities for recreation including swimming and windsurfing. (Photo: Eric Bird.)

cepted longshore drifting of sand, resulting in depletion of beaches down-drift. Beach sand has accumulated in Sandringham Harbour, alongside the breakwaters at Queenscliff and Mordialloc, and on a smaller scale in the harbours at St Leonards, St Kilda and Brighton, but elsewhere erosion has predominated. Attempts to restore depleted beaches have included bulldozing of sand from nearshore shallows south of Rosebud, a procedure which resulted in infestation of formerly clear sea water by seagrass vegetation and the accumulation of large quantities of drying weed on the beach. On several other sectors, depleted beaches have been artificially renourished with sand dredged from the sea floor and piped in to the shore (Figure 1). These have become popular recreational resources (Figure 2) and there are strong demands for much more extensive beach renourishment (Port Phillip Authority, 1977).

It has been predicted that man-induced changes in the Earth's atmosphere (the so-called Greenhouse Effect) will lead to global warming, and a world-wide sea level rise of the order of a metre over the next 60 to 140 years. Such a change would lead to further erosion of beaches around Port Phillip Bay, and would necessitate additional structural works to prevent marine flooding and erosion. More extensive renourishment will be required if recreational beaches are to be maintained in face of a rising sea level.

Recreational uses

Port Phillip Bay is intensively used for recreation by the people of Melbourne and

Figure 3. Recreational boats at anchor at Williamstown. The skyline of the city of Melbourne is in the background. (Photo: Eric Bird.)

Geelong, by those who live in bordering areas, particularly the Mornington Peninsula, and by visitors from other parts of Victoria, Australia, and overseas (Figure 3). The major recreational opportunities provided by Port Phillip Bay are both water-based (swimming, power boating, water skiing, fishing and sailing) and land-based (camping, picnicking, walking, sunbathing and other beach activities, and driving vehicles for pleasure). These activities need skilled planning and management if they are not to conflict with each other, and with the objective of maintaining the environmental resources of Port Phillip Bay. The problems of deciding how Port Phillip Bay and its coastal fringes are to be used are a matter of continuing media interest, political activity and aministrative debate.

Water-based recreation opportunities

Swimming. Beach use becomes intensive in the summer months and on a hot holiday afternoon there are up to 300 000 people on the beaches around Port Phillip Bay, with severe crowding on the more popular beaches, especially in the Melbourne area (Mercer, 1972). Conflicts between nearshore swimmers and those who wish to use surfboards, sailboards, dinghies and canoes are resolved by a longshore zoning scheme, and the use of power boats (e.g. for water skiing) is also restricted in the nearshore area. Most municipalities restrict the bringing of dogs on to the beach in the summer months, and there are a variety of by-laws concerning horse-riding, removal of shell fish, and sundry beach behaviour. Demands for nudist (naturist) beaches of the kind that exist in other states have been met on a limited scale by designating two sectors for this purpose, one on the west coast south of Altona, the other on the east coast north of Mornington.

Another problem with swimming opportunities comes from water quality. With large urban and industrial areas on its coastline, Port Phillip Bay receives substantial quantities of urban sewage and sullage and industrial waste, and problems of sea pollution have been evident, particularly in Corio Bay near Geelong and Hobsons Bay, at the mouth of the Yarra. In addition to outflow from rivers and creeks, there are about 300 storm-water drains discharging on to the shores of the bay. In recent years the Environment Protection Authority has monitored pollution in the bay, and has established controls and management systems that have succeeded in keeping it within tolerable limits. Concentrations of heavy metals (cadmium, lead, mercury) from industrial waste and garage effluent were high enough to restrict the harvesting of shellfish in parts of Port Phillip Bay in the nineteen-seventies, but these are now much reduced, as are concentrations of hydrocarbons and pesticides. Urban sewage is treated in large scale treatment systems: the broad plains south of Werribee on the west coast have been used to treat much of Melbourne's wet waste (134 million gallons/day) in the Melbourne and Metropolitan Board of Works Sewage Farm. Treated sewage is used here to irrigate pastureland, and then released into the bay, causing some eutrophication: concentrations of nitrogen compounds and phosphates are still rather high off the sewage outfalls. More recently, waste water from the south-eastern suburbs of Melbourne has been treated at the Carrum Downs processing plant, then piped into Bass Strait west of Cape Schanck. Sewage from Geelong has been treated and discharged into Bass Strait west of Barwon Heads.

The waters of Port Phillip Bay are discoloured by suspended sediment delivered in runoff during and after episodes of heavy rainfall over the bordering land areas. In calm weather a surface layer of relatively fresh turbid water may persist for several days, but usually the suspended sediment is quickly dispersed by waves and currents, and precipitated to the sea floor as the water becomes more saline. The effects of bacterial (*E. coli*) pollution from sewage discharged into coastal waters became a major concern of the Environment Protection Authority in the nineteen-seventies, because it was perceived to be a health hazard for people swimming in nearshore waters. Monitoring of *E. coli* in coastal waters soon identified the drains and creeks that were major sources of such pollution, and beach users were advised not to swim near these areas. In recent years this problem has diminished as a result of tighter controls on effluent discharge, and the recent development of more comprehensive and efficient sewage treatment systems.

Power Boating. Power boats are extensively used on Port Phillip Bay to support fishing activities, for water skiing, for underwater sports, and for pleasure boating. Interactions between these various activities, and with other water-based activities, have led to zonation of these activities in order to separate them.

Demands continue for the provision of more boat launching ramps and more moorings for recreational boats on Port Phillip Bay. Existing harbours and marinas have been extended and elaborated, in some cases resulting in the erosion of

Figure 4. St Kilda marina, showing structures built for dry storage of boats.
(Photo: Eric Bird.)

adjacent beaches. A new marina has been constructed at St Kilda (Figure 4), but proposals for additional marina construction have met strong local opposition, for example at Sorrento, where many residents object to the town's beach being enclosed as a harbour for recreational boat moorings. On the other hand, a new marina is under construction on the salt marshes bordering Swan Bay near Queenscliff. Demands for boat moorings have resulted in intensive development along the banks of estuarine creeks, as at Mordialloc and Patterson River, and the tidal channel at Queenscliff. A marina estate, with housing served by canal as well as road, has been developed alongside Patterson River.

Small power boats are launched from a number of boat ramps around Port Phillip Bay. There are demands for the construction and improvement of boat ramps, and for accompanying large car and trailer parking facilities, which have a major impact on the coastal area. The problem is exacerbated by the fact that local weather patterns include frequent late afternoon southerly winds, which can produce waves of up to a metre within less than half an hour. Boats that have been launched over an 8 hour period (e.g. between 5 a.m. and 1 p.m.) all have to be recovered from the bay in a short time under difficult sea conditions (Cullen, 1978).

Fishing. Marine angling is a major activity on Port Phillip Bay, with up to 1800 fishing boats operating at least once a year. Some shellfish (notably mussels and abalone) are harvested from rocky nearshore areas. Successive immigrant groups have locally over-cropped the nearshore resources, and it has recently been necessary to enforce strict controls on shellfish collecting by arrivals from southeast Asia, notably Vietnam, whose traditions permit more intensive exploitation of nearshore shellfish resources than is sustainable in Port Phillip Bay. Early in 1988 the situation was complicated by the development of a red algal bloom in the northern part of Port Phillip Bay, and consequent restrictions on the harvesting of shellfish which became toxic.

Port Phillip Bay has a substantial commercial fishery, including shell fish (notably scallops) dredged from the bay floor. Some areas have been allocated to mussel and oyster farming. Commercial fisheries generate about $A750 000 per

Figure 5. Summer camping in coastal *Blanksia* woodland at Rosebud, Port Phillip Bay. (Photo: Eric Bird.)

annum (about 15% of Victoria's annual fish production), the main fishing fleets being based at Mornington, Mordialloc, Frankston and Portarlington.

Sailing. Sailing is an activity that has been increasing, and changing its resource requirements in recent years. Port Phillip Bay provides an excellent environment for sailing for all levels from junior and social to international competition. Large boats require marinas and club houses on the coastal fringe, and these raise similar problems to those mentioned for power boats. Smaller boats and windsurfers launched directly from the beach cause occasional conflicts with beach users, especially swimmers.

Land-based recreation opportunities

Coastal recreation in Port Phillip Bay is facilitated by the presence of Crown Land reserves above high tide mark, typically at least 30 m wide, around the bay coastline. These were originally declared in the eighteen seventies, when only a few small sectors of the bay coastline had passed into private ownership (Bird, 1975). Subsequently, demands for a variety of uses have resulted in alienation of sectors of this public land, so that now about 11% of the land abutting the high tide line is privately owned, 44% under the jurisdiction of government departments and agencies (e.g. for port operations, navigation facilities, defence, sewage treatment), and 45% still coastal reserve.

Within the coastal reserve, however, there have been numerous developments and modifications. The need to service water-borne recreation has resulted in the establishment of about 40 boating clubs, each with facilities built on the coastal reserve for the exclusive use of club members. There are also about 40 other sports club facilities, including buildings for anglers clubs and sea scouts, as well as facilities for sports not related to water use, such as tennis, croquet, bowls and football. Areas of the coastal reserve have been developed as car parks and boat launching ramps, as picnic grounds, and as summer camping areas. Nevertheless, several sectors of the coastal reserve retain a bushland cover, forming a natural background to the beach and foreshore environment on a scale that is unusual near coastal cities. In general there is ready public access to the beach, the foreshore, and the sea. Along much of the east coast of the Bay a highway lies immediately behind the coastal reserve, and in a few sectors, notably between Mordialloc and Chelsea, housing development in the coastal reserve impedes public access to the shore.

Camping. Bayside camping has long been a popular summer activity, and some areas have been set up as extensive camp grounds under the *Banksia* and tea tree woodland that dominates the coastal fringe, especially in the southern part of Port Phillip Bay (Figure 5). The fees charged for camping generate substantial income for local authorities, much of which is used in the maintenance of camp grounds.

Walking and picnicking. Footpaths have been established in the coastal fringe for the benefit of people who wish to walk along sectors of the coastline of Port Phillip Bay, and areas have been set aside as picnic grounds, with tables and barbecue facilities.

Sunbathing and beach activities. The excellent beaches, mainly of sand, around the shores of Port Phillip Bay, and especially along the eastern coastline, attract up to 300 000 visitors on a hot summer day (Cullen, 1973a). Usage is heaviest in summer, between November and April, but the bay is used for various activities throughout the year.

Beach recreation requires the provision of public access to the shore, the availability of car parks, and service facilities such as life saving clubs, changing rooms and toilet and shower blocks. Sectors backed by dunes with a cover of scrub and woodland, as at Seaford, were damaged by trampling along numerous pathways from the highway to the beach, until the Port Phillip Authority, in association with the Soil Conservation Authority and the local municipal council, fenced the backshore area to permit regeneration of dune vegetation, restricting access to intervening walkways at intervals of about 100 m (Cullen, 1973b). In previous decades, boat sheds and bathing boxes were built on several sectors of the beach fringing Port Phillip Bay (Figure 6), and either owned or leased by holidaymakers, but these are now less fashionable, and most municipalities have reduced them. In 1977 there were still about 2000 such structures on beaches, the majority on the shores of the Mornington Peninsula. Some still consider that they add colour to the beach scene, and perhaps reduce erosion, but they have become controversial

Figure 6. Beach huts at Mount Eliza, Port Phillip Bay. (Photo: Eric Bird.)

because they alienate part of the diminishing beach area for use by a privileged few.

Driving for pleasure. Road access and car parking are significant problems, with overcrowding on peak usage days (generally 10 to 20 hot days in summer), but for the rest of the year (i.e. up to 355 days) the facilities provided are more than adequate. A number of car parks have been established in the coastal fringe, and there are occasional demands to extend these, or establish new car parks, at the expense of existing coastal reserves, but these have generally been resisted on the grounds that it is unnecessary to modify the coastal landscape simply to provide additional car parking when this is required only a few days per year.

Scenery and nature conservation. There is growing public interest in the maintenance of natural, or little modified coastal scenery and the conservation of features of scientific and historical interest (including nature reserves, bird sanctuaries, marine parks, archaeological sites and early settlement locations) in and around Port Phillip Bay. Some coastal municipalities have set up advisory panels on environmental management and published educational material to stimulate and enhance the appreciation and understanding of such features by local residents (e.g. Bird, 1983; Whiteway, 1985). Already established nature reserves include the Seaford Foreshore Reserve (Figure 7), Mud Islands, Edwards Point and the Sand Hummocks, a bird sanctuary. The southern part of Port Phillip Bay has been declared a Marine Park, and historical reserves include the 1803 settlement site at The Sisters, near Sorrento; the defence structures at Point Nepean and Queenscliff; and the man-made island at Pope's Eye. Local and state agencies concerned with the promotion of tourism in the Port Phillip Bay area are well

Figure 7. Coastal nature reserve at Seaford, Port Phillip Bay. (Photo: Eric Bird.)

aware of the need to establish and manage such reserves, and to provide interpretative material in the form of signage and publications.

Coastal management in Port Phillip Bay

The task of planning, organising and managing the use of Port Phillip Bay and its coastal fringes is the responsibility of the 18 municipalities which border Port Phillip Bay, and is administered by 39 Committees of Management. These receive government grants, and supplement their income by charging fees for car parking, boat launching, camping and caravanning in order to meet the costs of cleaning beaches and maintaining service facilities such as car parks, boat launching ramps and toilet blocks. Management is also complicated by the fact that about 20 state government agencies have responsibilities in the coastal area. These include the Ministry for Planning and Environment, the Department of Conservation, Forests and Lands, and the Port of Melbourne Authority.

In 1966 an attempt was made to co-ordinate planning, development, and management of the coastal area by setting up the Port Phillip Authority, responsible for the zone extending 200 m landward and 600 m seaward from the low tide line (Cullen, 1977). Demands for a wide variety of development, including marinas, boat launching ramps, car parks, sports club facilities, and restaurants then required the consent of the Port Phillip Authority, which initially consisted of a Chairman and representatives from the four major government agencies concerned with the coastal area. In the decade 1968–77, the Port Phillip Authority received 734 applications for consent to works or activities that would modify or impinge on the coastal environment, and some 90% of these were approved (Cullen, 1982). However, the Authority had limited powers and inadequate finance to deal effectively with the many and varied coastal management issues in the Port Phillip Bay area, and some (especially those disadvantaged by the Authority's requirements and decisions) criticised it as an unnecessary bureaucracy. The outcome was that, as the result of internal and external political pressure, the Authority was first restructured (in 1980), and then disbanded.

There is still no overall coastal management scheme for Port Phillip Bay. The competing demands are now resolved by government and agency decisions, sometimes with *ad hoc* public inquiries. Decision making is strongly influenced by local politics, some decisions depending more on political expediency, electoral manoeuvring, or media pressure than on scientific and economic investigation. This confused and disintegrated approach to coastal management has generated further dissatisfaction, and it is now widely acknowledged that planning is required to reconcile competing demands, and that the activities of the many participating agencies and organisations need to be co-ordinated, bearing in mind that activities or structural developments on one sector of the coastline may have adverse effects on other sectors. It is apparent that the existing management system for Port Phillip Bay divides control and creates tensions between state agencies involved with planning, land management, local government, sport and recreation, and tourism. Planning continues to be *ad hoc*, and land management is therefore of variable quality and effectiveness around the bay.

Opportunities in Port Phillip Bay

Australia, unlike many countries, has retained at least a strip of the coastal margin in public ownership (Cullen, 1987a). This gives unique opportunities, but also problems, as we have seen in Port Phillip Bay.

Managing public land, like planning, requires that the aim be clearly defined over specific time scales (next week, next year, next decade). Management planning is the process by which objectives are determined and appropriate strategies devised (Table 1). It traditionally includes a resource survey, an inventory of present usage and likely future pressures, and some attempt to estimate the carrying capacity of the land for the uses proposed. It assesses biophysical constraints, such as erosion and vegetation damage and proposes techniques to deal with them (Cullen, 1981).

The Recreational Opportunity Spectrum is the best available planning tool, aiming to provide a diverse range of sites in which various kinds of recreation may take place. Allied to this is the acceptance that there will be some degradation of these sites as they are used for recreation, the requirement that the manager specifies limits of acceptable change, and orgnises monitoring of the resource to determine when these limits are being approached. In the course of these activities, the manager will acquire an understanding of the ecological processes dominant on

Table 1. General principles of planning for coastal recreation.

1. The aim of recreation planning is to provide a range of sites that will enable people to find a situation that allows them to have the kind of experience they seek. In practice the best approach is the Recreational Opportunity Spectrum. This requires planners to provide a range of recreation sites, each of which is specified by explicit criteria such as size, access and maintenance costs. The spectrum works best with 6 to 8 types of site defined by key criteria. A corollary of the spectrum approach is that people need to know about the variety of sites if they are to realise the value of a recreation system. Each opportunity site is characterised by certain biophysical factors, and will have various management constraints if it is to be maintained.

2. In general, we do not have sufficient understanding of the recreation experience to provide background for decisions on planning or management. However, it is possible to recognise clusters or packages of recreational activities that commonly take place at a particular site on a particular visit: for example when a group of people visit a beach to sunbathe, picnic and swim.

3. User groups for various activities could be identified, and their potential size and distribution estimated, form socio-economic data and census information, but this approach has not yet been used in recreation planning in Australia.

4. Each opportunity site has a certain capacity to tolerate use, beyond which there will be unacceptable impacts, either on the site (e.g. exposure of bare soil and tree roots, track erosion, and litter and waste accumulation) or nearby (bacteria and sedimentation in nearshore water). Managers need to determine acceptable limits of usage and change at a particular site.

5. The key management tool in protecting natural areas is control of access through the provision of roads, tracks, car parks and signs. Permits or booking may be used to limit the numbers using a particular site. Without adequate control of access, other management tools are not effective.

6. Managers may need to control access over short periods of perhaps a few months to allow regeneration of sites where the impacts are becoming unacceptable.

7. Once planning has determined the type of opportunity site to be provided, design is required to ensure that suitable facilities are available.

8. In reserves, interpretation of the resource (by providing signs, information boards, pamphlets etc.) is a key to shaping visitor behaviour towards protection of the resource.

a particular site, and the opportunities and constraints on its use. Effective management planning requires the manager to work in a team with planners and resource specialists during the planning phase. Too often the managers are excluded from the planning process, which is carried out by consultants who may lack detailed local knowledge, and who do not have to live with the long-term results of their work (Cullen, 1987b). Management plans are of little use without adequate management staff on the ground: people who know, understand and care for the resource. Coastal areas are rarely staffed adequately.

It is hoped that existing efforts by state and local government agencies to devise guidelines for coastal management in Port Phillip Bay will result in sufficient staff and funding to be able to plan and manage the bay's intricate resources effectively. It is also necessary to provide an educational programme which will show local people the consequences of unwise planning and development, such as the building of houses too close to eroding beaches, or the pollution of beaches and coastal waters where sewage and sullage are discharged on to the shore. It is necessary to avoid the situation where politicians accept the claim that a proposed development will benefit the local economy and provide employment at the cost of only minor environmental degradation of the coastal area: such developments are apt to proceed sequentially, and lead to cumulative degradation of the coastal environment.

The present fragmented system of control of the use of the resources of Port Phillip Bay is unwieldy, and prevents effective planning and management. If the recreational use of a system as complex as Port Phillip Bay is to proceed without environmental degradation, these organisational structures must be improved.

References

Bird, E. C. F. 1974, Man's impact on the Melbourne coast. *Victoria's Resource*, **16**(3), 12–15.
Bird, E. C. F. 1975, The management of coastal reserves. *Victoria's Resources* **17** (2), 3–7.
Bird, E. C. F. 1983, *Geology and landforms of Beach Park: an excursion guide*. Sandringham Environment Series, 2.
Bird, E. C. F. 1985, *Coastline changes*. Wiley Interscience, Chichester.
Cullen, P. 1973a, *Recreational demands on the beaches of Port Phillip Bay*. Port Phillip Authority, Melbourne.
Cullen, P. 1973b, Coastal conservation problems at Seaford. *Victorian Naturalist* **90**, 4–9.
Cullen, P. 1977, Coastal management in Port Phillip. *Coastal Zone. Management Journal* **3**, 291–305.
Cullen, P. 1978, Planning for boating facilities on the coast, *Australian Parks and Recreation* (November 1978), 35–44.
Cullen, P. 1981, *Managing Coastal Resources*. R.A.I.P.R. Management Manual Series, Canberra.
Cullen, P. 1982, Coastal zone management in Australia. *Coastal Zone Management Journal* **10**, 183–212.
Cullen, P. 1987a, Managing the coastal heritage of Austalasia, *Coastal Zone 87*, 461–463.
Cullen, P. 1987b, Coastal resource management and planning, *Australian Planner* **25**(3), 10–12.
Mercer, D. C. 1972, Beach usage in the Melbourne region. *Australian Geographer* **12**(2), 123–139.
Port Phillip Authority 1977, *Port Phillip Coastal Study*, Melbourne.
Whiteway, B. 1985, *Marine life of the coastal fringe*. Sandringham Environment Series, 6.

Eric Bird
Department of Geography,
University of Melbourne,
Victoria,
Australia

Peter Cullen
School of Applied Science,
Canberra College of Advanced Education,
Canberra,
Australia

4. Recreation in the coastal areas of Singapore

Introduction

The Republic of Singapore consists of one main island of 570 km² and 57 offshore islands, the largest of which is 17.9 km². The coastline is 136 km long for the main island and more than 300 km if the offshore islands are included. 'If the coastal areas are taken to mean a strip two miles wide along the foreshore (excluding all rivers), this strip occupies more than 50% of the main island and all the offshore islands' (U.N. Department of International Economic and Social Affairs, 1982).

For this paper, the coastal area is defined more narrowly in relation to its recreational use. The coastal area includes the nearshore water and the beaches, where a variety of recreational activities can be carried out, and adjacent public parks that have facilities not necessarily related to the beach or seawater. Although the coast on Singapore Island is accessible within an hour from any inland location, the idea of staying at seaside accommodation appeals to the average Singaporean who lives in a flat. Holiday bungalows and chalets built specially to take advantage of the seaside environment are therefore included in this paper. Strictly speaking, the local population staying more than 24 h in such accommodation would be considered as domestic tourists. Coastal recreation in this paper is therefore taken to encompass all recreational activities including holidaying carried out in the coastal areas.

The objectives of this paper are 1) to describe the present spatial patterns of coastal recreation; 2) to examine the major developments in coastal recreation; and 3) to discuss some of the problems and issues associated with coastal recreation in Singapore.

Spatial patterns of coastal recreation

Despite a favourable ratio of coastline to area (1 km/1.9 km²), coastal recreation is not widespread but localized to specific stretches and offshore islands. Coastal recreation is virtually excluded from the western half of the main island where the land is reserved for military purposes, reservoirs, industries, and port operations. It is confined predominantly to the eastern third of the main island and offshore islands zoned for recreation. This pattern coincides largely with short stretches of natural beaches or beaches that have developed on the reclaimed coast of the main island and the offshore islands.

The basic patterns of coastal recreation remain as in the 1:50 000 map on recreation (Mapping Unit, 1981). The major changes had been the loss of Pulau Tekong to military use and Pulau Serangoon for reclamation, and several new

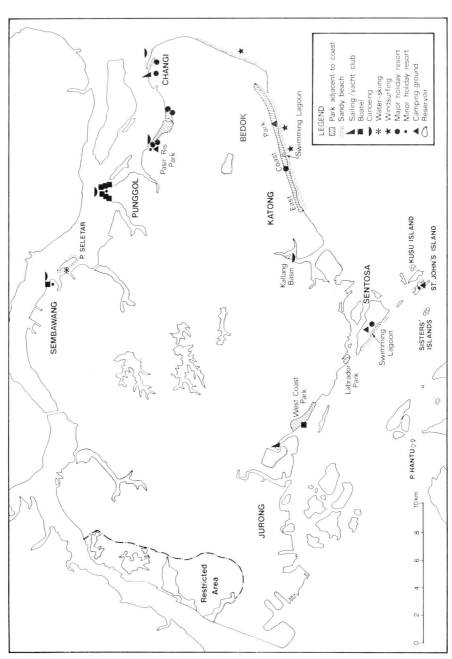

Figure 1. Coastal recreation in Singapore.

recreational developments on the main island and the offshore islands. Figure 1 shows the patterns of coastal recreation at the end of 1987. A brief description of the major areas of coastal recreation is given below.

East Coast Park. The East Coast Park is built on about 140 ha of reclaimed land and has 7 km of beaches from Bedok to Katong. Its nearshore area is used for swimming, boating, sailing, windsurfing and fishing. The park has a man-made lagoon which was opened for swimming in May 1976. Windsurfing is offered by two organizations: a private company, which was established in 1980 and uses part of the swimming lagoon to train windsurfers; and a sea sports centre, which was set up in 1987 by the People's Association (PA), the government's premier community development agency. Two kilometres east of the lagoon is a campsite which can accommodate about 500 campers. More permanent accommodation is found in the East Coast Park Chalets complex consisting of 110 one-storeyed units and 59 two-storeyed units. The chalets average an occupancy rate of 85%, as a result of their yearly leases to companies and their popularity during the weekends and school holidays (*Straits Times*, 19.12.86). Other facilities in the park include cycling tracks, jogging tracks, a golf driving range, a tennis centre, an aquatic complex, and a seaside restaurant centre. These were developed in the 1970s with private participation encouraged by the government (Wong, 1983). The East Coast Park, with its wide range of facilities and easy accessibility, has grown to be the most popular stretch for coastal recreation in Singapore.

Offshore islands. Sentosa is the most popular of the offshore islands for recreation, catering to both local population and tourists. It is linked to the main island by a ferry service and a cable car. Coastal recreation is centred on the swimming lagoon, the boating lagoon and the adjacent Siloso Beach and Tanjong Beach, all of which are being marketed as part of the 'sun world', one of the four 'worlds' on the island. Twenty-six holiday bungalows are leased only to corporations for at least a year. The island is continuously developed by the Sentosa Development Corporation, a statutory body, to attract more visitors.

The Sentosa Development Corporation also manages twelve other offshore islands for the government: Kusu Island, St. John's Island, Sisters' Islands (Pulau Subar Laut and Pulau Subar Darat), Pulau Hantu, Pulau Seletar, Terembu Retan Laut, Buran Darat, Pulau Jong, Pulau Biola, Pulau Renget, and Lazarus Island. The first five islands are popular for swimming, sailing, scuba-diving, snorkelling and fishing. Kusu Island and St. John's Island are served by regular ferry. The former island is particularly popular during the ninth lunar month when pilgrimages are made to the island's temple. Four holiday bungalows and a holiday camp are also available on St John's Island.

Changi. As a result of land taken up for Changi Airport in the mid-seventies, the Changi-Tanah Merah coast lost its popularity. With recent improvement to facilities however, the remaining stretch of beach and park between Tanjong Changi and Changi Creek is regaining popularity. West of Changi Creek the beaches are poorer but the area remains an important coastal recreational zone, centred on the government holiday bungalows, a sailing club and a swimming club. Changi Village, which was redeveloped in 1976, is an added attraction for the holiday crowds and boasts the only hotel in Singapore with immediate access to a beach.

Pasir Ris-Loyang. The Pasir Ris-Loyang coast has developed into the major holiday resort area in Singapore despite land reclamation resulting in the loss of Pasir Ris beach and the Pasir Ris Beach Hotel. The holiday bungalows at Loyang were the earliest, some of which still remain. In November 1973, the Pasir Ris Holiday Flats, facing the then Pasir Ris beach, were completed. Immediately west of these, an eight-unit resort belonging to Telecoms was built. To the east, with improved accessibility and the completion of a private housing estate, many of the unsold bungalows facing the sea were eventually bought by companies and statutory boards for holiday purposes. Between 1978 and 1980, 44 ha east of the park up to Sungei Loyang were reclaimed and protected by a series of breakwaters. This reclaimed land forms part of the 80-ha Pasir Ris Park to be developed in three phases, of which phase I at the western end, has been completed. On the reclaimed land at Loyang, a group of 37 holiday units were completed in November 1984 for civil servants. To the west of it and also on reclaimed land is the Pasir Ris Resort which opened in December 1987 and consists of 94 single-storeyed units and 92 double-storeyed units. Further west is the 396-room resort which belongs to the Singapore Labour Foundation and is expected to be completed in mid-1988.

Punggol. For more than 30 years Punggol has served as an important centre for boatels. Six boatels provide storage spaces for about 1000 speed boats and cabin cruisers, look after bigger cruisers anchored at sea, and rent boats for water skiing, fishing and sightseeing. All are affected by the Punggol reclamation project and have been allowed to operate for another five years (*Straits Times*, 24.3.86). Punggol has a few holiday bungalows and a holiday camp.

Sembawang. At the end of Sembawang Road is a small park, Sembawang End Park, and a short stretch of beach which is popular for swimming and fishing. Further to the east is a boatel, a canoe club and holiday camp. West of Sembawang Shipyard is the SAF (Singapore Armed Forces) Yacht Club.

Kallang Basin. The 2.8 km newly created beaches along the Kallang Basin banks of the Marina Centre, and along the banks of the Rochor and Kallang rivers, are the latest addition to the beaches of Singapore. These culminate the efforts of the Ministry of the Environment in its programme to clean the Singapore River/Kallang Basin catchments since 1978. A sea sports centre is located below Merdeka Bridge.

Other areas. With an area of 50 ha on reclaimed land, the West Coast Park, which was completed in 1985, is the second largest park after the East Coast Park. Due to few facilities, the park is less popular. The beaches are narrow and pebbly in character and are used for a temporary boat station. The reclaimed coast parallel to

Changi Coast Road has well formed beaches but the area is not developed and access is limited. The waters off its southern half are used for windsurfing.

Major developments in coastal recreation

The demand for recreation increased considerably in recent years as a result of an increasing population and, more important, rising expectations from a more affluent, urbanized and high-rise living population. For example between 1976 and 1986, per capita indigenous GDP increased from $5496 to $13 088 and the percentage of population living in public flats increased from 60% to 85% *(Yearbook of Statistics, 1987)*. This demand has translated into a need for more land for recreation with coastal recreation becoming increasingly important. The more important developments are identified below.

Move to the reclaimed coast. The Seventies saw the emergence of coastal recreation along stretches of reclaimed land and on some reclaimed islands. After the success of its pilot project at Bedok in 1963, the Housing and Development Board embarked on the large scale reclamation of the east coast which was protected by series of breakwaters between which beaches formed (Wong, 1981). A decade of facilities development has made the east coast the most popular stretch for coastal recreation.

In the Seventies the Port of Singapore Authority also reclaimed and enlarged many of the southern islands that were sparsely populated and under-utilized. Sandy and shelly material were placed behind stone bunds constructed on the reef flat and beaches with seaward convex shapes formed between these structures (Wong, 1985). With their new beaches, Pulau Hantu, Pulau Subar Darat, Pulau Subar Laut, Kusu Island and St. John's Island were opened to the public in 1975.

Island resort. Following the withdrawal of British forces in 1967, the government decided in 1968 to develop Sentosa, which was formerly known as Pulau Blakang Mati, into a holiday resort to attract more visitors to Singapore. Dillingham Overseas Corporation *et al.* (1970) carried the feasibility study and the development of Sentosa into an island resort was undertaken by the Sentosa Development Corporation, which was formed in September 1972. Among the projects, those related to coastal recreation were the swimming lagoon (ready by 1974), a canoeing lagoon (1975), a campsite (1975), holiday chalets and a youth hostel (1977) for large organized groups *(Mirror*, 15.10.82). In 1980 the Pacific Area Travel Association (1980) was engaged to reappraise the developments on Sentosa. A resort hotel operated from late 1978 to March 1987. Subsequently, it was let out as chalets and closed in December 1987 to make way for a new resort hotel with 275 rooms. Other proposed developments related to coastal recreation include a coralarium and sea life park, medium-class hotels and campsites *(Straits Times,* 1.12.87). Sentosa has evolved into one of the popular areas for recreation for the local population and tourists.

Increased range of coastal recreation. Until recent years, Singaporeans regarded the sea as a place for swimming and water-skiing. In developing a comprehensive recreational programme, the People's Association acted as a catalyst through various activity clubs to popularize and increase the range of recreational activities along the coast (People's Association, 1985). Thus, canoeing and board sailing (windsurfing) have become very popular in recent years. In 1987, three sea sports centres were opened at the Kallang Basin, Pasir Ris, and the East Coast in addition to three existing clubs catering mainly to canoeing at Changi, Punggol Point and Sembawang.

New patterns in seaside holidaying. Civil servants have had in the past the possibility to apply for seaside holiday bungalows and chalets at a number of seaside locations. Due to coastal reclamation, such units closed down at Tanah Merah, Tanjong Rhu and Pasir Panjang, leaving the 40 odd units at Changi. These were supplemented by those at Loyang. Previously, the allocation, which is handled by the Ministry of Finance, was on a weekly basis. This was changed to two periods per week (Monday p.m.–Friday a.m.; Friday p.m.–Monday a.m.) in July 1985 to encourage better utilization. Although the rentals have also increased since November 1984, they are low compared with commercial rentals (*Sunday Times*, 25.11.84). Several statutory boards own or lease holiday bungalows at various locations on the main island and offshore islands.

Holiday accommodation for the general public was initially provided by the People's Association, which operated several holiday camps. Those currently available are at Punggol Point, Kampong Tengah (Sembawang) and Pasir Ris and they are very popular during the weekends and school holidays. The People's Association also maintains a 4-ha campsite at the East Coast Park. The Pasir Ris Holiday Flats were developed to cater for its own staff and the public in the lower income brackets (*PA Annual Report, 1969*). Together with the holiday camp, they form the Pasir Ris holiday complex. As expected, the peak usage coincides with school or public holidays.

The demand for seaside holiday bungalows by the public led to the construction of the first group of such accommodation at the East Coast Park by the Urban Development and Management Company, a government private limited company, in December 1977. The success of the first phase of single-storeyed units led to the implementation of the next phase of double-storeyed units in October 1981. The company also developed another resort at Pasir Ris in 1987. Other holiday accommodation available to the public include the bungalows and a holiday camp on St. John's Island and the hotel at Changi. Workers will have their own resort at Pasir Ris with the completion of the largest holiday resort in Singapore by the Singapore Labour Foundation.

Problems and issues

As it is largely based on water and the beach, coastal recreation has specific problems, e.g. threat from oil pollution, poor beach formation, inadequate facilities for boats. The issues it faces are those related mainly to competing uses of the coast and some of these issues are still outstanding.

Oil pollution. Despite regulations and precautions, oil pollution poses a constant threat to coastal recreation. Some 600 ships are moored or pass through Singapore's harbour every day. The port handles more than 70 million tonnes of oil and several refineries are located on the west coast and offshore islands. Between 1971 and 1987, eleven major oil spills, with five in 1987, have occurred. Between them, they have polluted the beaches along Sembawang, Changi, Bedok, the east coast, Marina Bay and Sentosa. The most common cause of small-scale pollution is contributed by bunkering or oil cargo loading, ballasting and de-ballasting operations. Although spills are more likely to occur in the western sector of the port waters, oil pollution is not confined to any specific area (*Straits Times*, 9.10.87). Many of the methods used to fight oil spills make use of a chemical dispersant to break the oil (*Straits Times*, 25.7.87).

Poor beaches. The two artificial swimming lagoons at the east coast and Sentosa have maintenance problems. The East Coast Swimming Lagoon has to be desilted annually because of siltation and pollution from the organic waste of cerethid shells. In the Sentosa Swimming Lagoon the pollution from cerethid shells has recently been controlled by regular pumping of the lagoon's water. As a result of continuous trampling on the beach by picnickers and erosion from surface runoff, the beach gradient is progressively lowered. Beach improvements based on wave tank studies are being implemented.

Due to low wave energy, the beaches along the reclaimed Pasir Ris-Loyang coast are poorly formed as compared to the beaches along the east coast. Mangrove plants have also re-colonized the reclaimed coast. Narrow beaches characterize the coastal stretch of government bungalows at Changi.

Insufficient facilities for boats and sea ports. The need for more facilities for boats and sea sports has been a perennial issue in coastal recreation. An example is the joint letter sent to the Prime Minister by five associations representing the sailors, skiers, power boaters and windsurfers. In the letter the associations pointed out the absence of a government policy to sea sports. They wanted the government to allocate coastal land zoned for sea sports facilities as an increasing number of Singaporeans were taking up sea sports. They also noted that there was no modern marina facility in Singapore. The government's response was that 'setting land aside for boating facilities was low priority because they generate little or no revenue' and suggested a possible site at Sentosa, which was limited to planning a marina for sailing dinghies (*Straits Times*, 18.4.84).

At present more than 4000 pleasure craft are registered with the Port of Singapore Authority and these are distributed among the Republic of Singapore Yacht Club (RSYC) at Sungei Pandan, the SAF Yacht Club at Sembawang, Changi Sailing Club, and the boatels mainly at Punggol and Sembawang. Storage for canoes is available at Sembawang, Punggol Point, Pasir Ris, Changi, and Kallang Basin.

Over the years, proposals for marina development at Punggol to replace the boatels had little success. This became clear when the government announced the plan for the Punggol reclamation in April 1985. The boatels were affected; one had moved out and the remaining six were given another five years stay. The proposed marina development by five of the present boatels will only be considered after reclamation is over. Meanwhile the Ministry of National Development has earmarked two plots at the west coast and Pasir Ris for marinas (*Straits Times*, 6.10.86). The west coast site is offered to the RSYC as its present site is restricted by the Pandan bridge; the Pasir Ris site is to be developed by Singapore Power Boat Association and Singapore Water Ski Federation (*Straits Times*, 2.1.87).

Public vs private interests. In the past many private residences along the coast at Punggol, Loyang, Tanah Merah, Bedok, Siglap, Katong, Tanjong Rhu and Pasir Panjang enjoyed direct access to the sea, limiting public accessibility to the beaches. Reclamation has resulted in the loss of sea frontage for nearly all these residences, and the beaches on the reclaimed coast are now accessible to the public. One objective of the government is to open the coast to public use where possible and restrict development for private use. Thus, it rejected a hotel's proposal to develop a stretch of Changi beach for its guests in 1985 and specify that facilities to be developed on URA (Urban Redevelopment Authority) sites must be open to the public and not for members only (*Business Times*, 26.1.87, 13.3.87). As early as 1964 the Foreshore Ordinance was passed to disallow claims by owners for the loss of their sea frontage.

The government has also taken measures in the interests of safety in coastal recreation. Windsurfing has been banned in the East Johor Strait which is a busy shipping channel. To reduce the number of boating accidents, drivers of speedboats and other powered pleasure craft have to undergo a test and be licensed from April 1987 (*Straits Times*, 10.4.87). New safety measures have been taken after an accident in January 1983 when the top of a drillship collided with the Sentosa cableway resulting in the loss of seven lives. Over the years it becomes apparent that the legislative efforts of the government in relation to coastal recreation have led to a drastic reduction of private ownership of coastal areas and islands but increased access to the beaches and safety to users.

Planning for coastal recreation. In national planning, the coastal areas are not considered as a separate entity. Also, there is no single national organization responsible for the coast. The Port of Singapore Authority has jurisdiction over the waters and land within the port limits; pollution on beaches and in water courses are dealt with by the Ministry of Environment; parks including those bordering the coast are the responsibility of the Parks and Recreation Department; Sentosa and other offshore islands zoned for recreation are under the Sentosa Development Corporation; and other specific coastal areas related to recreation are under different organizations. At the national level, recreational activities including those at the coast are the responsibility of the People's Association and the Singapore Sports Council. Private clubs and private organizations involved in coastal recreational activities are confined mainly to boatels, sailing, wind surfing clubs and such like.

The coastal areas and some of the offshore islands will play an increasing role in future recreational development. The proposed provision of open space including private open space and recreational areas in the latest revised Master Plan is 3303 ha in 1985 or about 4.7% of total area of Singapore. The objective is to achieve an overall standard of 0.8 ha per 1000 population for public open space; this is expected to be achieved by 1990 and assuming a population of 2.75 million by 1990, the area of public open space required would be 2200 ha. Since 1980, the revised Master Plan has included West Coast Park, Pasir Ris Park and Kent Ridge Park, the first two being near the coast. Future park development exists in Marina City, the offshore islands and the Comprehensive Development Areas (Planning Department, 1985). The islands zoned for recreation include Sentosa, Buran Darat, Pulau Serangoon, Kusu Island, Lazarus Island, Pulau Renget, Pulau Jong, Sisters Islands and Pulau Hantu (Planning Department, 1985). Due to the property slump, commercial development has been held over for the 300-ha reclaimed land at Marina South. Instead an interim land use plan (20-year) to provide more recreational and leisure facilities for the public in the 1990s is being implemented (*URA Annual Report, 1985–86*).

Aspects of coastal recreation are also included in other development plans at the national level. Although not singled out specifically, recreational facilities along the coast have been implemented by the Singapore Sports Council in the first phase of the Master Plan of Sports Facilities, 1976–82 in which development was guided by a number of factors (Singapore Sport Council, 1983). In the recently released tourism development plan for Singapore (Ministry of Trade and Industry *et al.*, 1986), the proposals to develop Singapore as a tropical island resort, have a direct bearing on coastal recreation. These proposals include projects on Sentosa, Lazarus Island and marinas and are meant not only for tourists but also for the local population.

Conclusion

The demand for coastal recreation is increasing and this can be met by the improvement of existing facilities or the provision of new facilities and services. Except for Sentosa, the other offshore islands zoned for recreation require improved or new facilities and services. Modern marinas are also required. Beach improvement is recommended at Pasir Ris-Loyang, Sentosa and at the coast fronting the government bungalows at Changi. Due to increasing demand, there is scope for more imaginatively planned holiday resorts. For planning purposes, a more integrated perspective to coastal recreation is recommended as existing facilities and services are under different public and private organizations.

Acknowledgements

The author wishes to thank the officers in charge of recreation in various public and private organizations for their prompt responses to his inquiries.

References

Business Times, 26 January 1987; 13 March 1987.
Dillingham Overseas Corporation *et al.* (1970), *A Resort/Recreation Plan for Blakang Mati, Singapore.*
Mapping Unit (1981), *1:50 000 Recreation Map of Singapore*, Singapore.
Mirror, 15 October 1982.
Ministry of Trade and Industry *et al.* (1986), *Tourism Product Development Plan*, Singapore.
People's Association (PA) *Annual Report*, 1969, 23.
People's Association (1985), *25 Years with the People 1960–1985*, Singapore, 112.
Pacific Area Travel Association (1980), *A Development Review of Sentosa*, San Francisco.
Planning Department (1985), *Revised Master Plan 1985 Report of Survey*, Singapore, 51–54, 63.
Singapore Sports Council (1983), *The First Ten Years*, Singapore, 75.
Straits Times, 18 April 1984; 24 March, 6 October, 19 December 1986; 2 January, 10 April, 25 July, 9 October, 1 December 1987.
Sunday Times, 25 November 1984.
United Nations Department of International Economic and Social Affairs (1982), *Coastal Area Management and Development*, Pergamon, Oxford, 74.
Urban Redevelopment Authority (URA) *Annual Report*, 1985–86, 8–9.
Wong, P. P. (1981), 'Beach evolution between headland breakwaters', *Shore and Beach* **49**, (3), 3–12.
Wong, P. P. (1985), 'Singapore', in E. C. F. Bird and M. L. Schwartz (eds.) *The World's Coastline*, Van Nostrand Reinhold, New York, 797–801.
Wong, Y. K. (1983), 'The greening of Singapore: An introduction to parks and recreation works' in O. W. C. Fung and S. Y. Ng (eds.), *Recreation and the Community*, Institute of Parks and Recreation, Singapore, 17–21.
Yearbook of Statistics, 1987, Department of Statistics, Singapore, 2, 18.

P.P. Wong
Department of Geography,
National University of Singapore

5. The Azov Sea coast as a recreational area

The Azov Sea and its coast possess considerable recreational resources. The Azov health resorts are scattered in a vast territory in the Krasnodar, Rostov, Donetsk, Zaporozhje, Herson and Crimea regions. Recreational development of the Azov Sea coast started later than that of the Black Sea and Baltic Sea because of its less impressive landscapes. Natural woods are completely absent, monotonous relief predominates, coastal plain abruptly falls to the sea forming abrasive shores, with narrow beaches along the scarp. These and other conditions (limited water resources, inadequate accessibility) resulted in inefficient utilization of the Azov Sea recreational potential. In the early seventies, the situation started to change especially in the regions adjoining industrial centers of Donetsk basin, Azov region and lower Don river. Now the Azov resort zone hosts 900 000 tourists a year. Organized recreational inflow exceeds 400 000 a year, 50% of which stay at recreation centers. The majority of holidaymakers (75%) accumulate in the northwestern sector, where recreational industry is represented by small uncontrolled recreational services of local nature. Development of organized recreation is concentrated on the northern sea coast and in Taganrog Bay. Remote areas, such as Arabatskaya spit, northern coast of Kerch peninsula, Kuban river sea coast and the eastern part of the water basin, are still poorly developed.

Intensive recreational development of the Black Sea coast in the summer results in concentration of holiday-makers on small areas, causing difficulties in meeting sanitary standards. Development of new territorial recreation systems and improvement of transport access allowing to involve new areas, can help to obviate to this problem. The Azov Sea coast should become one of these new recreational zones in the nearest future, providing a wide network of sanatoriums, boarding houses, campings and children's institutions for a total of one million residents. A favourable climate, considerable resources of various mineral waters and mud-baths contribute to this decision.

The Azov coast climate is favourable for tourism and holiday-making from the end of April till the end of October. The percentage of warm, sunny days is as high as 70%. The duration of sea-bathing season is about four and a half months: from 127 days in the northern part up to 138 days on the Krimean coast of the Azov Sea. Summer temperature of the sea water near the beach is up to 31° C. In winter the sea is covered with ice for 1–3 months. The water is curative and is used for rhodonous, oxigenous, hydrosulfide, nitric baths; and also for gargling, inhalations, spraying.

Curative muds play an important role in the development of health-resorts on the Azov Sea coast. They are represented mostly by various silty sulfide sediments. The most famous are the muds of Krasnoye, Chokrakskoye, Sivashskoye, Golubitzkoye and Khauskoye deposits. Sea-mud deposits with immense operational resources have been found in Taganrog and Berdyansk Bays. A peculiar type of curative muds (volcanic muds, formed during an eruption of thin pelitic matter mixed with mineral iodobromine waters from the mud-volcano vents) is found on the Taman (Gnilaya, Kuchugary hills) and Kerch (Bulgakansk, Dzhau-Tepe) peninsulas.

The Azov Sea coast is also rich in hydrothermal resources. Chloride, sulfide, chloride-sulfide, hydrocarbonate, hydrocarbonate-sulfide and thermal waters are found here. It should be noted that more than 15 water sources can be found on the northern coast alone; these waters are mainly chloride-sodium and calcium-sodium ones. Mineral waters are available in Salgir and Chokrak resorts: iodobromine and sulfide waters in the former, carbonic acid and hydrosulfide waters in the latter.

Among natural conditions favourable for recreation, geomorphology plays an important role. The Azov Sea coast is plain, of liman type. It is composed of Neogene-Quaternary loose sediments, with gully-ravine and valley ruggedness. Complex geomorphology is the result of differences in tectonic conditions, rock composition, hydrodynamics, and such specific process as an intensive biogenous accumulation.

The specific feature of the water basin geomorphology is attributed to numerous accumulative forms alternating with the bedrock shore sections composed of sandy-clayey material. In terms of recreational potential, accumulative forms – spits, bars and points – are the most favourable. In the regions adjoining large industrial centers, abrasion and abrasion-slump shores are being developed.

The largest tectonic unit in the geomorphological zoning, suggested by Mamykina and Khrustalev (1980), is the coastal zone, distinguished by the tectonic setting and history of evolution. It includes regions differing in rock lithology of shore scarps, dominant shore type and lithodynamic environment. Subdivision of the region into separate areas is based on certain combinations of shore types, interdependence and intensity of their development. Using this scheme of zoning, it becomes possible to determine the regions and areas most favourable for recreational development in terms of geomorphology. While evaluating the possibility of their development the authors paid special attention to the occurrence of shore types (abrasion, abrasion-slump, accumulative), their dynamics, beach types, sediment composition, relief and sediment of the subaqueous shore slope.

The most developed in terms of economy is the northern coast of the Taganrog Bay. Areas in the eastern part of the bay are most densely populated. Industrial and residential buildings of the bedrock shore extend from the bench of the coastal scarp for 1,2 km in villages and up to 10 kms in towns. Within cities and villages the bedrock coast, the beach or accumulative terrace and the submerged slope, that is three major components of the coastal zone, are used. Industrial buildings are usually very near to the scarp bench and sometimes occupy slump terraces which has a negative impact on the coastal zone development, where the abrasion rate reaches 1–2 m yr. The extent of beaches and a small thickness of sediment is due to deficit of coarse-grained material in the coastal zone. The small width of the zone (from 2 to 10 m) makes their use limited.

In terms of recreation the coast of the Bay is insufficiently developed. Priority is given to easily accessible sites adjoining settlements with easy slopes as well as large accumulative forms. Thus, on the south-eastern shore of the Krivaya spit, 11 km long, numerous recreational buildings are erected. The width of the spit beach increases from 5–7 m at the near-root end up to 25–30 m at the distal end. Sediments are dominated by quartz sands, detrital limestone, pebble, gravel; on the spit tip, shelly content increases. The sea floor slope is 0.02–0.27 at the central portion of the south-eastern shore; 0.003 – at the spit tip, while the subaqueous slope of the north-western shore is more shallow (0.002). Recurrent westerly storms accompanied by the water rises cause spit erosion. This process intensifies at the sites with shore-protective structures, such as concrete walls and slopes, impeding accumulation processes at sea coasts. The Krivaya spit has a considerable recreational potential if properly used.

Also promising as an accumulative form is the Beglitzkaya spit. Its direction was formed under the influence of westerly swell and sediment transport. The spit is 5 km long and 3 km wide. Elevation above the sea level is 0.5 to 1.5 m. The beaches 5 to 25 m wide are composed of quartz sand, pebble, gravel and shell. Gentle slopes and shallow water are favourable for children resorts. Silty bottom sediments decrease recreational value of the region to some extent.

In the future it will be possible to build up artificial beaches and to expand natural ones by addition of bottom sandy material. The 'free' beach in Taganrog city, 50–300 m wide and nearly 3 km long, is the first of this type, built up in 1975–1980. Its condition and dynamics demand regular artificial nourishment for stability maintenance.

The southern coast of the Taganrog Bay has favourable natural conditions, with wide sandy beaches which have sediment inflows due to erosion of the shore scarps. The beaches, composed of well-sorted sand with minor shells, have a width of 60 m at the mouths of large gullies and ravines. Their recreational development is hampered by common occurrence of abrasion-slump shores with cliffs as high as 30–45 m and narrow crests. Other negative aspects of natural conditions are shallowness of the submerged slope (with gradient as low as 0.001–0.005) and a high amplitude of water level fluctuations, especially in the eastern part of the bay (more than 5 m).

The southern shore of the Bay is characterized by attached accumulative forms, similar in morphology and dynamics, having the shape of asymmetric projections with predominantly unilateral feeding. These are Ochakovskaya, Chumburskaya, Sazalnitzkaya spits 2–3 km long and 3–7 wide at the base. The flat near-root parts of these spits, having an elevation of 1–1.5 m above the water line, can be used for construction only if the surface is raised by an artificial fill. The recreational centers are being built now near the beaches 6 m wide in the root part and 60 m – in the distal part. The western shore has slopes of 0.012–0.04 and the eastern slope – 0.07–0.09; beach sediment is composed of quartz sand with minor shells and detrital materials (20–30%) and pebbles (5–15%).

Other spits of the southern shore (Yeyskaya, Naydennaya) do not have any recreational value because of small dimensions. The city beach in Yeysk can be expanded at the spit base by filling of shore-protecting interjetty pockets.

The most suitable recreational zones of the northern sea coast are the unique accumulative forms, known in literature as 'the Azov-type spits': Berdyanskaya,

Obitochnaya, Fedotos spit – Island Biruchy, and Belosarayskaya spit, situated at the boundary of the Taganrog Bay. Their distal parts are declared sanctuaries, and the island of Biruchy is a reserve: any construction is forbidden here. Spit lengths increase to the south-west from 15 km (Belosarayskaya) to 46 km (Fedotov spit – Island Biruchy). The flat triangular basement of the accumulative forms is formed by the system of ancient bars and hollows, the elevation increases from 1–2 m in the south-western direction up to 5 m on the Island Biruchy. Their south-eastern shores are flanked by the beach attached to the foredune or shore bar. The western shores facing the bay have an irregular shape, with limans, bays and islands. Their elevation is from 0.5–1.0 to 5 m, the gradient up to the depth of 5 m is 0.002 to 0.008. During surges the spit is partly flooded so that construction on the spit base and western shore is possible only after artificial raising of the surface has been performed.

The beach width of the spits increases from 5–15 m in the root part to 100–150 m in the distal part. A special feature of sediment composition is the decrease of terrigenous and the increase of biogenous components, the process being directed from the root part to the distal part, and from Belosarayskaya spit to Biruchy Island for the coast as a whole. The shell content increases along the spit contour from 2 to 80%, and from Belosarayskaya spit to Biruchy island – from 20 to 90%. Beach gradients vary from 0.04 to 0.15 on the sea shore of the spits and from 0.01 to 0.45 on the distal end. The subaqueous slope reaching the depth of 2 m has a gradient varying from 0.01 to 0.50.

Spit dynamics should be taken into account when laying out recreational zones. Repeated survey of the outer contours of the accumulative forms showed seasonal changes in lithodynamic processes. In winter, with easterly swells, erosion dominates the root part and accumulation takes place at the distal end, while between them there is a relatively stable portion of sediment transit. In summer, with westery swells and surges, sediment redeposition occurs along the outer contour from the distal end to the spit base. The amount of the sediment transported as well as the beach width depend on the swell direction and energy parameters.

When describing accumulative forms, one should note a favourable aesthetic impression, with wide shelly beaches stretching for tens of kilometres, shallow bays, purity and warmness of sea water.

Construction of recreational centers is in progress on the abrasion and abrasion-slump shores. The site adjoining the Belosaraysk and Berdyansk bays is most intensively developed in spite of unfavourable geomorphology: elevations up to 68 m high in the Zhdanov region, sharp scarps with the rate of erosion as high as 4 m yr, small beach widths (2–20 m) and silty sediments on the subaqueous slope. Local distribution and small thickness of the beaches are caused by the deficit of beach forming fractions in the abrasive material. Small rivers of the northern coast – former sources of alluvial sediments – have practically ceased to provide them.

Increasing rates of shore development necessitate expansion of the available beaches, construction of artificial ones, and the use of rational methods of shore protection. Inwash of 'free' beaches is possible on a number of sites with limited migration, particularly on the rear spit sides of the northern sea coast and Taganrog Bay at the junction with bedrock shore.

The Arabatskaya spit – the largest accumulative form of the Azov sea (110 km long, 0.3–5 km wide) – is the most promising in terms of recreation. It represents a bar, or an enormous barrier, separating the sea from the Sivash limans. The lower surface with ancient bars parallel to the modern shore line and alternating hollows is situated behind the wide beaches. The beach elevations are 3 to 5 m above the sea level, their width is 20–30 m in the northern part and 50–60 m in the central and southern parts. Corresponding gradients are 0.02–0.45 and 0.006–0.15.

The Arabatskaya spit is composed mostly of shell and detrital material (80–85%) with 20% of sediment represented by quartz sands in the northern parts. The biogenous material comes from shell banks and results from erosion of ancient sediments. The submerged slope is deep and has a shell bar at the depth of 2–2.5 m. Its upper part has a gradient of 0.01–0.1 and 0.003–0.006 at the depth of 5 m. A considerable portion of the Arabatskaya spit shore is subject to erosion.

Withdrawal of shell material resulted in sediment deficiency, especially during the periods of low productivity of zoobenthos.

In spite of favourable natural conditions (shelly beaches, steep coast, clear water, free water exchange and good aeration), free spaces for building between the bars and the Sivash bays are still inadequately used.

The negative factors in the western part are high surges (up to 3 m) during easterly storms.

The Kerch-Taman coast and the Kuban delta shore have a considerable recreational potential. The coastal zone of the Kerch peninsula is poorly developed. The recreation centres and boarding houses are situated in the vicinity of Kamenskoye, Mysovoye, Peschanoye and other villages on rather low shores and bars of the Kazantypsky bay, Aktashckoye and Chokrakskoye lakes, etc. The picturesque eastern part of the peninsula with rock cliffs 10–20 m high and lime points forming small bays with local sandy and pebble beaches 20–25 m wide is practically undeveloped. Sandy sediments with minor shell cover the steep subaqueous slope with gradient of 0.01–0.03. Two larger bays – Reef and Bulganak, 8 and 9 km long with beaches 10–30 m wide, gentle bottom slopes and 2–3 subaqueous barriers – are of special interest in the long-term planning of recreational development.

The bar of the Kuban delta sea edge, from the Achuevskaya spit to Akhtanizovsky liman, more than 150 km long, composed of quartz sand, shell and detrital sediment, borders the alluvial-deltaic plain with inlets, channels and limans used for hunting and fish breeding. The resort development is limited to the narrow zone (100–200 m) with 10–50 m wide beaches of the Akhtanizovsky liman with the slope gradient of 0.06–0.1. The submerged slope with 3 to 4 barriers has a gradient of 0.02–0.003 to the depth of 3 m. The development of the delta sea edge is complicated by shallowness of the shore and recurrent surges during northerly stormy winds. Disastrous surges of 1969, exceeding the bar height, resulted in bar's erosion and deltaic area flooding.

The eastern section of the coast has favourable natural conditions and is extensively used for agriculture. The surface of the coastal plain is completely plowed up, but poorly developed as a recreational area. The health resorts are adjoining the cities of Yeysk, Prymorsko-Akhtarsk and Dolgaya spit. They are planned for further development.

The geomorphology is rather specific here: the abrasion shores dominate, cliffs are 7–11 m and up to 18 m high near Yeysk. Narrow beaches (5–10 m) at the scarp base are composed mainly of coarse-grained material with little sand and silt. The beach has a gradient of 0.012–0.05 with the subaqueous slope dip of 0.02. The Prymorsko-Akhtarsk region has the largest shore erosion rate – up to 6 m a year, and 4 m yr near the Yeysk which is caused by high westerly swell and surge levels.

The accumulative forms of the shore, the Dolgaya and Kamyshevatskaya spits, are the newly developed health resorts. There are holiday and boarding houses on the Dolgaya spit. This accumulative form stretching for nearly 30 km in the north-west, separates the Taganrog bay from the sea and determines its hydrology and hydrochemistry. Its width is 6 km at the base and up to 500 m at the distal end. The maximum width of beaches on the south-western sea shore is 40 m, and the north-eastern shore – 10 m; the corresponding gradients are 0.015–0.20 and 0.01–0.016. The beach sediments are mostly shells. 1 metre water depth is 300 m away from the bay water line and 200 m from the sea water line; the gradients of the subaqueous spit slopes is 0.003–0.006 and 0.005–0.008 correspondingly. The distal end of the Dolgaya spit can change its slope due to hydrodynamics and the amount of the biogenous material transported form the open sea banks and the bay floor. Thus, only the relatively stable central part between the spit base and the distal end, and the base of the accumulative form are favourable for recreational development.

It can be concluded that the recreational potential of the Azov sea coast is sufficient for extensive development of holiday resorts, but is rather poorly used now. The regions with the favourable natural conditions, the Arabatskaya spit, the bases of the northern shore and Taganrog bay spits, should be given priority and the Kerch-Taman coast and the eastern coast should come next. To eliminate the negative impact of the recreational stress on the coastal zone, the comprehensive development of the territory is necessary with regard to specific features of the coast as a whole and of each individual site.

V.A. Mamykina and Yuri P. Khrustalev
Geological-Geographical Department,
Rostov State University,
U.S.S.R.

6. The influence of ethnicity on recreational uses of coastal areas in Guyana

Introduction

Guyana, formerly named British Guiana, is a culturally-pluralistic society with six distinct ethnic groups. With the exception of the Amerindians who live in the sparsely settled interior region, ninety percent of Guyana's 900 000 people is concentrated on a narrow coastal zone which is less than five percent of the country's land area. The population of the coastal zone or coastal belt reaches 400 per km^2, whereas the country's overall population density is only 4.7/km^2.

The residential preference is matched by a preference for recreation in the coastal zone by all ethnic groups whose members are descendants of culture groups from Africa, Asia, and Europe.[1] The primary objective of this paper is to, therefore, investigate whether there are differences in recreational use of the coast by the various ethnic groups which, especially in the semi-urban and rural areas, live in distinct communities. Since settlement of the coastal zone is a consequence of European expansionism and colonization it becomes necessary to provide a brief background on the geographical and historical setting of Guyana.

Guyana – location and physiography

Guyana is located above the equator in the northern coast of South America between 0°41' N and 8°33' N and 56°32' W and 61°22' W. With an area of 214 970 km^2, Guyana is bounded by the Atlantic Ocean, Brazil, Surinam and Venezuela (Figure 1). The country can be divided into four physiographic regions: the Pakaraima mountains, the Pre-Cambrian lowlands, the sandy, rolling lands, and the coastal plain which is the most used for recreational activities (Figure 2).

The coastal plain – geography and settlement

The coastal plain of Guyana is approximately 1.5 m below sea level at high tide and extends a distance of about 435 km from Punta Playa in the west to the Corentyne River in the east. Varying in width of between 77 km in the west and 26 km in the east, the coastal plain occupies roughly eight percent of Guyana's land area (see Figure 2). Of significance, is the fact that of Guyana's population of about 900 000, nearly ninety percent live on the narrow coastal plain. Most people use the resources of the coastal plain, as for example, the various types of beaches (Figure 3) for recreational purposes. A comprehensive description of beaches and other physical aspects of the coastal plain can be found in Vann (1959) and Daniel (1984).

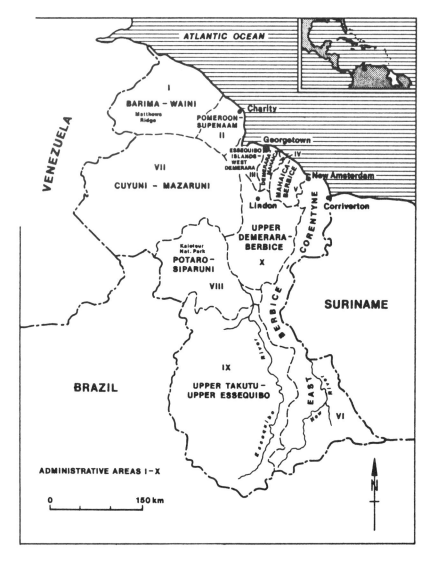

Figure 1. Location of Guyana.

Macpherson (1969) correctly pointed out that ever since the first successful cultivation of the coastlands began, virtually all further settlement has grown up there, and the interior has remained unaltered.[2] The coastal plain has always been attractive for settlement partly because of its fertile alluvial soils, and also because of a year round growing period with average annual temperature of 27° C and a mean annual rainfall of 2.3 m.

The settlement of Guyana's coastal zone is a direct result of European expansion in the Caribbean. Details on Guyana's colonial history can be found in several studies (for example, Rodway, 1891; Newman, 1964; Daly, 1974; Shahabuddeen,

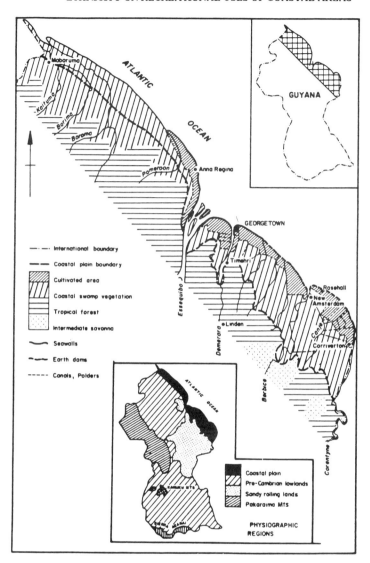

Figure 2. Physiography and the coastal plain.

1978). In brief, the Dutch transported slaves from Africa during the seventeenth century to work on the cotton, tobacco and sugar plantations on the coastlands. In 1812, five years after the abolition of the slave trade, the African slave population in Guyana was estimated at 100 000 (Jeffrey and Baber, 1986). The British acquired Guyana from the Dutch in 1814, and when the British ended slavery in 1833, they had to find other sources of labour to work on the plantations. The alternative sources which were used included Portuguese from Madeira, Chinese, and East Indians from India. Only the last of these sources proved satisfactory, and consequently from 1838 until 1918 (when the system of indentured Indian im-

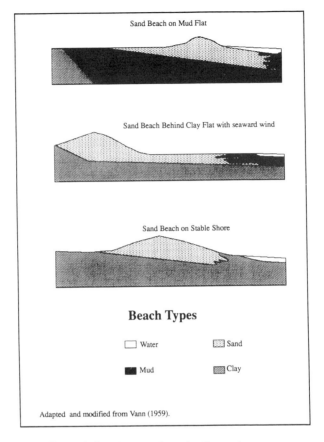

Figure 3. Beach types along the Guyana's coast.

migrants ended) there were about 239 000 Indian immigrants, seventy percent of all indentured immigrants (D. Nath as cited by Milne, 1981).

Ethnic composition and cultural pluralism

Today, Guyana's population reflects its history of immigration and colonialism, because the British colonial system executed a deliberate policy of using differences in ethnic groups and cultures to divide and rule (see Williams, 1957; Adamson, 1972). Owing to its diverse ethnic composition (Figure 4), Guyana is often called the 'land of six races'. Given the fact that 'the populations that came to Guyana from Africa, Asia and Europe differed not only phenotypically but also in their cultural traditions and practices' (Despres, 1975), several writers have referred to the Guyanese society as being socially and culturally pluralistic. While the plural society concept has been widely debated (see e.g., Smith, 1960; Despres, 1967; McKenzie, 1967; Cross, 1971; Nicholls, 1974) the author of this study, neverthe-

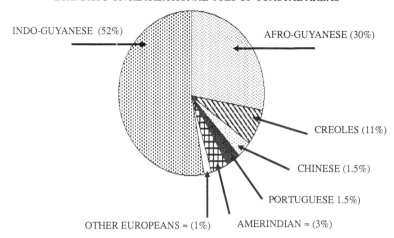

Figure 4. Ethnic composition of Guyana's population.

less, considers Guyana to be a pluralistic society because each ethnic group represents a distinct and competitive sociocultural entity.

Study objectives and design

Since each ethnic group in Guyana has retained distinctive aspects of its own identity, this study aims to:
a) find out how the coast is used for recreational activities by the various ethnic groups,
b) determine whether the ethnic groups in rural, semi-urban and urban areas engage in different coastal recreational activities, and
c) investigate whether there are differences in recreational use of the coast by the various ethnic groups.

To fulfill the objectives of this study, three different coastal locations representing urban (Kitty), semi-urban (Ogle-Plaisance communities), and rural (Fyrish-Gibraltar communities) have been intensively studied (Figure 5). These communities have been selected because of their ethnic composition, and also because each community can take advantage of the widest range of coastal recreational activities. Kitty is located on the outskirts of the capital city of Georgetown, and has an ethnically mixed (Indo-Guyanese, Afro-Guyanese, Chinese, Portuguese[3] and Creoles[4]) population of more than 10 000. The semi-urban adjacent communities of Ogle and Plaisance are located approximately 10 km from Georgetown, and have respective concentrations of Indo-Guyanese and Afro-Guyanese. The rural communities are about 125 km from Georgetown, with Fyrish having about 1100 Indo-Guyanese and Gibraltar about 1000 Afro-Guyanese.

To supplement a total of more than twelve years of interviews and observations the author made while working and living in the urban, semi-urban and rural communities, this study also collected in 1983 and 1984 seven hundred and twenty field questionnaires by employing random, and stratified random sampling

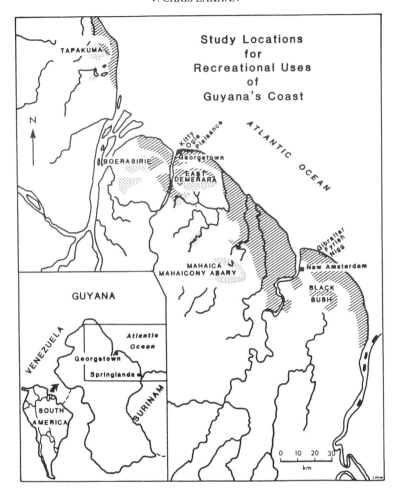

Figure 5. Study locations for recreational uses of Guyana's coast.

techniques. Based on the number of adult members in each community, 416 questionnaires were obtained from Kitty, 100 from Ogle, 96 from Plaisance, 58 from Fyrish and 50 from Gibraltar. Each questionnaire collected information on several study related attributes, among them ethnicity, primary coastal recreational activity, frequency of primary and other recreational activities, time spent, and time when engaged in recreational activity, and socioeconomic and demographic characteristics.

Overview of observations and survey results

Irrespective of ethnicity and social and economic status, the majority of Guyana's population use the coast for one or more forms of recreational activities. There are,

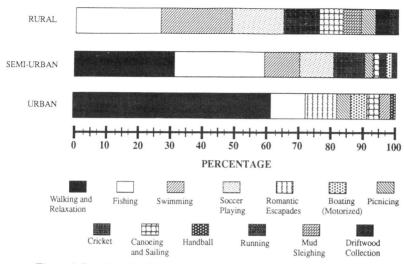

Figure 6. Rural, semi-urban and urban recreational uses of Guyana's coast.

however, preferences for certain types of recreational and leisure activities, and from the graphical and statistical results the conclusion can be made that there are distinct spatial and ethnic variations in recreational uses of Guyana's coast.

Figure 6 demonstrates the noticeable differences in recreational activity between rural and urban areas. The bar graph clearly shows that the majority of people in urban and semi-urban areas enjoy the coast for walking and relaxation while the primary recreation activity of people in the rural communities is fishing. It is also demonstrated that a combined total of 37.9% of rural dwellers engage in swimming (22.3%) and soccer playing (15.6%), while in the urban areas no one plays soccer[5], but 3.8% of the people use the coast for recreational swimming. Of interest here is the fact that although the semi-urban communities of Ogle and Plaisance are less than 8 km from the urban area of Kitty, yet 11% of people engage in swimming, and another 11% play soccer.

While several reasons can be cited as to why the urban residents of Kitty prefer only certain types of recreational activities, the one that is of paramount importance is the finding that the ethnically-mixed population of Kitty has developed a 'seawall subculture'.[6] The residents of Kitty have a 'love affair' and preoccupation with the seawall (see Figure 2) which was constructed to protect the below sea level coastlands around Kitty from flooding. Details on the seawall and sea defence strategies can be found in Daniel (1988).

At all times of the day, and during the year, people of all ethnic groups either walk or sit and relax on the concrete seawall, and enjoy the waves and the invigorating sea breeze. As in the case of recreation in the Australian coastal zone (see Yapp, 1986), 'the benefits of "sea air" and, in particular, the relief afforded by the cooling sea breezes' have influenced the recreational habits of urbanized Kitty, and neighbouring Georgetown. With regularity, government, business and office workers, especially the unmarried, 'flock' to the seawall after work to walk and

KITTY - URBAN

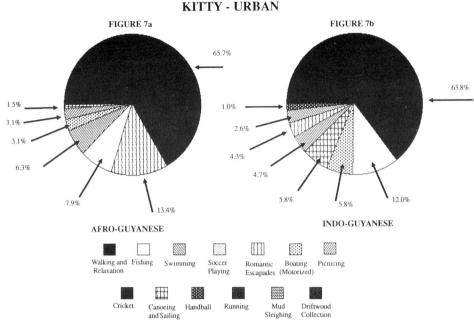

Figures 7a and b. Urban Afro-Guyanese and Indo-Guyanese recreational activities.

KITTY - URBAN

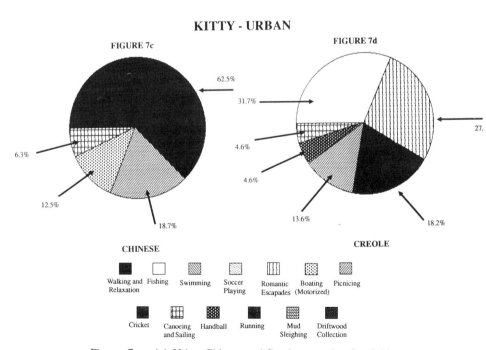

Figures 7c and d. Urban Chinese and Creole recreational activities.

ETHNICITY ON RECREATIONAL USES OF COASTAL AREAS 77

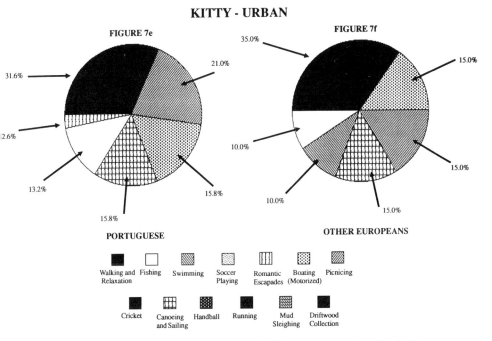

Figures 7e and f. Urban Portuguese and other Europeans recreational activities.

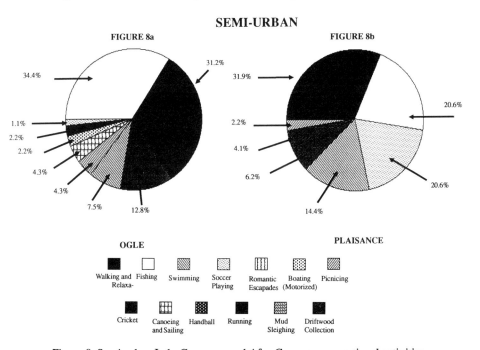

Figure 8. Semi-urban Indo-Guyanese and Afro-Guyanese recreational activities.

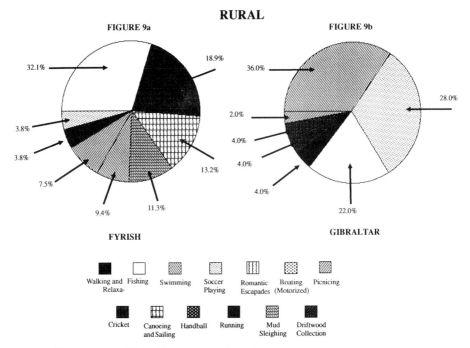

Figure 9. Rural Indo-Guyanese and Afro-Guyanese recreational activities.

relax. Figures 7a to 7f clearly show that of the various ethnic groups more than sixty percent of Afro-Guyanese, Indo-Guyanese and Chinese in the urban area enjoy walking and relaxation as a recreational activity. What cannot be demonstrated by these figures is the fact that there is very little interaction between and among these ethnic groups although they participate in the same recreational activity. This author concurs with Despres (1975, p. 105) who observed that, '...at the seawall where people gather to enjoy the evening breezes, ...it will be observed that Africans and Indians sit or walk separately and they do not frequently greet one another or join in conversation'.[7]

A further examination of Figures 7a–7f reveals that different ethnic groups have greater preferences for certain types of recreational activities. For instance, it can be observed that 13.4% of Creoles (Figure 7d) consider romantic escapades as a primary recreational activity, while only an extremely small percentage of the other ethnic groups indulge in this activity. Interestingly, those who use the seawall to engage in romantic escapades are frowned upon by members from the other ethnic groups. It is not uncommon to hear Indo-Guyanese and Chinese referring to Afro-Guyanese and Creoles who enjoy romantic escapades as 'white niggers', and 'redskin bitches'.[8] Without doubt, the Indo-Guyanese and Chinese resent the fact that the Creoles and some Afro-Guyanese have been acculturized to certain European norms and values. What is surprising is that wealthy Indo-Guyanese and Chinese are proud when they can pursue motorized boating (see Figures 7b and 7c) as done by the Portuguese (Figure 7e) and other Europeans (Figure 7f).

Although economic mobility has allowed superficial interaction between rich members of the various ethnic groups, it must be stressed that each ethnic group considers itself to be unique in Guyanese society. To the inexperienced 'outsider' a member of any ethnic group will portray himself/herself as a Guyanese first and foremost. However, once the outsider becomes fully acquainted with that member of any ethnic group, then his/her 'Indian-ness', or 'Black-ness', or 'Chinese-ness', or 'White-ness' will be fully understood. This tendency and intrinsic preoccupation with ethnicity in Guyanese society is further highlighted by examining the results dealing with the recreational behaviour of the semi-urban and rural communities.

Without reproducing the Chi-Square Contingency Tables and test results, the conclusion can be made that,

a) recreational uses of the coast by the Indo-Guyanese community of Ogle differs from that of the adjacent Afro-Guyanese community of Plaisance, and

b) recreational uses of the coast by the Indo-Guyanese community of Fyrish differs significantly from recreational uses of the coast by the adjacent Afro-Guyanese community of Gibraltar.

From the pie graphs (Figures 8a and 8b) it could be clearly seen that the Indo-Guyanese in Ogle (Figure 8a) and the Afro-Guyanese in Plaisance (Figure 8b) have preferences for certain types of recreational and leisure activities. The Indo-Guyanese use the coast for a diverse number of recreational activities, while the Afro-Guyanese concentrate mainly on walking, fishing, soccer playing, and swimming. Interestingly, the Indo-Guyanese of rural Fyrish also use the coast for more and different types of recreational pursuits than the Afro-Guyanese community in adjacent Gibraltar.

From Figure 9b it is seen that a total of 86% of the Afro-Guyanese of Gibraltar are engaged in only three activities, namely, swimming (36.0%), soccer playing (28.0%) and fishing (22.0%). On the other hand, a total of only 45.3% of the Indo-Guyanese of Fyrish (Figure 9a) fish (32.1%), swim (9.4%) and play soccer (3.8%). Although the two communities share the same socio-economic characteristics, the Indo-Guyanese of Fyrish are 'not contented and relaxed' like the Afro-Guyanese of Gibraltar. As a result, with imagination and innovation, the Indo-Guyanese of Fyrish pursue a diverse range of recreational activities. For example, mud sleighing is a recreational activity unique to Fyrish. Interestingly, while Indo-Guyanese especially those of rural Fyrish, view the beach and coastal waters as sanctified areas, some of their recreational pursuits, (for example, cricket and mud sleighing) cause the removal of accretion building sand and mud from the beach and coastal foreshore environments.

Conclusion

More than ninety percent of Guyana's population live in the coastal zone, and this residential preference is matched by a preference for recreation in the coastal environment by all ethnic groups, except the indigenous Amerindians who live in the country's interior areas. The results of this paper have shown that with cultural pluralism whereby each ethnic group tries to maintain distinct cultural traditions and practices, preferences have been developed for certain types of recreational activities in the coastal zone. Both the graphical and statistical results demonstrate that the ethnic groups in rural, semi-urban and urban communities engage in different coastal recreation activities, and that there are differences in recreational use of the coast by the various ethnic groups. It is evident that ethnicity not only controls Guyana's politics (see Hope, 1985; Jeffrey and Baber, 1986), but also influences recreational behaviour. This finding contradicts the observation by West and Heatwole (1979) who found that in New York City beach recreation is influenced more by economic and social conditions rather than ethnic background. With the knowledge that there are spatial and ethnic variations in recreational uses of Guyana's coast, recreation and allied planners can now make sound management decisions on the use and allocation of resources to promote recreation in a dynamic coastal environment.

Notes

1. Historical settlement of Guyana's coast required huge investments of capital and labour. It can be argued that with the exception of the Europeans, and Amerindians, the other ethnic groups were forced by the plantation owners in the eighteenth and nineteenth centuries to settle in specific areas along the coast.
2. Attempts by the State, especially since 1966, to develop and populate the hinterland areas of Guyana have met with only little success.
3. The Portuguese in Guyana have always been distinguished from Europeans. According to R.T. Smith (1962, p. 102), 'the circumstances of their arrival in the country, and the roles they came to play resulted in their acquiring a special identity'.
4. 'Creole' is a French word which has been adopted into the West Indian vocabulary. It is derived from the Spanish 'criollo', meaning 'native to the locality'. It was originally used by the South American Negroes and Spaniards to distinguish their own children from Negroes and Spaniards freshly arrived from Africa and Spain (Jayawardena, 1963, p. 3). In Guyana, the meaning of 'creole' has been extended to refer to 'mixed' people or 'coloured' people. This author found that people detest being called 'mixed' or 'coloured'. If an individual is called 'coloured' or 'mixed', the response will be 'mixed with what' or 'are you not also coloured'?
5. It would be misleading to suggest that soccer is not played in the urban areas. Soccer is played as a form of organized sport on the numerous cricket and soccer playing fields.
6. Long-term observations indicate that the cycles of erosion and accretion along Guyana's coast govern the intensity and frequency of seawall use. With an accretional cycle, associated with the growth of mangroves and other vegetation in front of the seawall, use of the seawall for recreational purposes declines.
7. This author observed that some co-workers, acquaintances, etc. will nod when passing each other. However, several indicators show that in the last decade, ethnic conflict has been exacerbated by state policies and political institutions.
8. Ethnic stereotyping, namecalling, group slandering, etc. is an every-day occurrence in an increasingly intolerant ethnically heterogeneous Guyanese society.
9. As members of each ethnic group become older, the differences in recreational uses of the coast by each group become more distinguishable.

References

Adamson, A.H., 1972. *Sugar without Slaves. The Political Economy of British Guiana, 1838–1904*. Yale University Press, New Haven.
Cross, M., 1971. On Conflict, Race Relations and the Theory of the Plural Society. *Race* **12**, 477–494.
Daly, V. T., 1974. *The Making of Guyana*. Macmillan, London.
Daniel, J. R. K., 1984. *Geomorphology of Guyana*. Occasional Paper No. 6, Department of Geography, University of Guyana. Georgetown, Guyana.
Daniel, J. R. K., 1988. Sea Defence Strategies and their Impact on a Coast Subject to Cyclic Pattern of Erosion and Accretion. To be published in *Journal of Ocean and Shoreline Management*, Amsterdam, Holland.
Despres, L. A., 1967. *Cultural Pluralism and Nationalist Politics in British Guiana*. Rand McNally and Company, Chicago.
Despres, L. A., 1975. Ethnicity and Resource Competition in Guyana Society. In: Despres, L. A. (ed.), *Ethnicity and Resource Competition in Plural Societies*. Monitor Publishers. The Hague, 87–117.
Hope, K. R., 1985. *Guyana: Politics and Development in an Emergent Socialist State*. Mosaic Press, Oakville, Ontario.
Jayawardena, C., 1963. *Conflict and Solidarity in a Guianese Plantation*. The Athlone Press, University of London, London.
Jeffrey, H. B. and Baber, C., 1986. *Guyana: Politics, Economics and Society*. Frances Pinter Publishers, London.
Macpherson, J., 1969. *Caribbean Lands. A Geography of the West Indies*. Longmans, London.
McKenzie, H. I., 1967. The Plural Society Debate: Some Comments on a Recent Contribution. *Social and Economic Studies* **15**, 53–60.
Milne, R. S., 1981. *Politics in Ethnically Bipolar States. Guyana, Malaysia, Fiji*. University of British Columbia Press, Vancouver.
Newman, P., 1964. *British Guiana: Problems of Cohesion in an Immigrant Society*. Oxford University Press, London.
Nicholls, D., 1974. *Three Varieties of Pluralism*. St. Martin's Press, New York.
Rodway, J., 1893. *History of British Guiana*. Vol. I, 1668–1781. Vol. II, 1782–1833. J. Thomson, Georgetown, Demerara, Guyana.
Shahabuddeen, M., 1978. *Constitutional Development in Guyana, 1621–1978*. Guyana Printers, Georgetown, Guyana.
Smith, M. G., 1960. Social and Cultural Pluralism. In Rubin, V. (ed.), *Social and Cultural Pluralism in the Caribbean*. Annals of the New York Academy of Sciences, New York, 736–767.
Smith, R. T., 1962. *British Guiana*. Oxford University Press, London.
Vann, J.H., 1959. The Geomorphology of the Guiana Coast. In *Proceedings, Second Coastal Geography Conference*. Held on April 6–9, 1959 at the Coastal Studies Institute, Louisiana State University, Louisiana, 153–187.
West, N., and Heatwole, C., 1979. Urban Beach Use: Ethnic Background and Socio-environmental Attitudes. *Resource Allocation Issues in the Coastal Environment*. The Coastal Society, Arlington, Virginia, 195–207.
Williams, E., 1957. The Historical Background of Race Relations in the Caribbean. In *Miscelanea de estudios dedicados a Fernando Oritz*. Havana, Cuba.
Yapp, G.A., 1986. Aspects of Population, Recreation and Management of the Australian Coastal Zone. *Coastal Zone Management Journal* **14**, 47–66.

V. Chris Lakhan
Department of Geography,
University of Windsor,
Canada

7. Recreational uses in the coastal zone of central Chile

This paper discusses the outstanding features of recreational uses of the central Chilean coast. Along this coast, stretching in a north to south direction, the resources of landscape and climate make possible tourist activity; likewise the vicinity of the principal urban centers of the country make these shores recipient of an important part of the domestic tourist demand.

New means of transportation and improvement of roads, in the middle of the 19th century, became important factors in the upsurge of public recreation. The construction of the Santiago-Valparaíso railway in 1854, linked the urban centers near Valparaíso. According to Mendez (1987), the beaches of Valparaíso, between 1860 and 1880, became tourist zones. The fashion of the sea bath transformed the ancient villages of fishermen into wonderful cities: Viña del Mar, for instance, gradually became a tourist town, and the same happened with the farms of Concón, Quintero, Zapallar and Papudo, towards the north and in the beaches of the province of Santiago, from Algarrobo, Cartagena up to San Antonio (see map).

The environment

The climate of Central Chile Coast is mild warm with winter rain (csb) and long dry season in summer. The winter rain is produced by frontal systems with an annual average of 350 mm. This cyclonical type of rains is affected by the coastal range, and for this reason, the annual amount of rain is larger in the littoral than inland.

The dry season lasts seven to eight months and it is produced by an uninterrupted anticyclonic dominion, with frequent overcast skies.

The thermometric state is influenced by the proximity of the sea and the annual average thermal difference is 6.5 °C. The average temperature in summer is 17 °C to 18 °C and 9 °C to 10 °C in the winter. There is a strong predominance of south-west winds, responsible for the formation of important dune fields.

The main morphological feature corresponds to the coastal range that stands out to the west littoral plains placed between the ocean and the main dividing range. The coastal range rises up to 2000 m at the latitude of Santiago. Its average width is 50 km and it has been an obstacle for communications between the littoral and the interior.

The coastal plains are vast flatlands arranged at regular intervals from the west bottom of the coastal range to the ocean. At the latitude of Santiago, there is a plateau of 200 m of altitude, but this has been lifted by tectonic movements and near Valparaíso it is 250 m height. These plains of marine erosion, are constructed in plutonic rocks of the coastal batholith. The plains' surfaces are intensely weathered and also eroded by river streams that dissect them (see map).

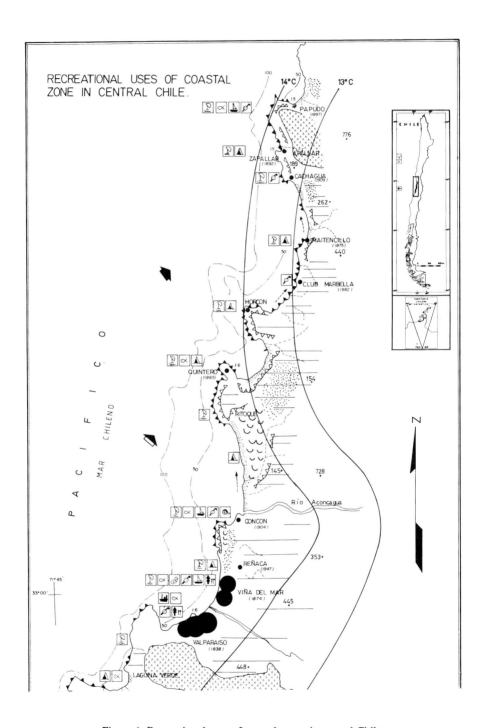

Figure 1. Recreational uses of coastal zones in central Chile.

RECREATIONAL USES IN THE COASTAL ZONE OF CENTRAL CHILE

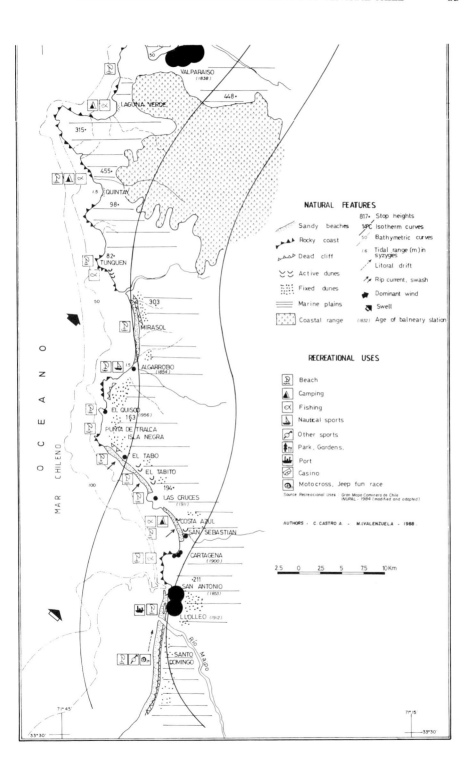

Recreational uses of the coastline

Coastal section Papudo-Maitencillo. The coastal range is next to the littoral (Papudo-Zapallar). The coastline is rocky with many reefs and low erosional platforms. The occidental hillsides of this mountain range are cut by many intermittent streams in which there are characteristic granitic boules, produced by the washout of weathering materials.

The beach resorts of Papudo and Zapallar are located in protected bays. At present it is observed the growing development of the linear urbanization around the coast route that connects these beach resorts and also between Zapallar and the area of dunes at the south of Cachagua.

Coastal section Maitencillo-Aconcagua River. The coastal plains meet an extensive development and there are ancient dunes, naturally fixed by vegetation.

Next to the beach resort of Maitencillo is Marbella Club, recently built (1982), located on the high plain, without direct contact to the sea. This fancy private club has a swimming pool, tennis and golf courts open to the public in week-days; furthermore it has a Club House and a discotheque which are the center of Zapallar and Cachagua night life.

Between Ventanas and Loncura, an industrial zone (a power plant and a copper refinery) on the coastline interferes with the tourist use of this sector, which is one of the most popular centers of tourism of the littoral.

The most extended beach of this zone, Ritoque, is exposed to the southwest wind and behind it there are active transversal dunes. Actually there are some summer houses at its north end and organized campings can be found at the south end of the beach.

Coastal section Concon-Viña del Mar-Valparaíso. This is the most popular area for tourism in Chile and in summer the most crowded zone of the whole central coast; Concón-Reñaca, Viña del Mar and Valparaíso constitute a great unit of business and recreation for tourism.

The littoral route runs 30 km between Concón and Valparaíso, bordering almost in its whole extension small bays, rocky headlands and beaches.

Between Concón and Reñaca, the coastal plains, 80 m height, are connected to the shore through a steep rocky cliff. The urbanization is intense in the rough slope cliff and on the dune fields over the coast plain.

Viña del Mar is set up as the most important tourist city of Chile, because it has all the necessary services for tourist activity. It is also called the 'Garden City' because it is surrounded by large green areas.

Valparaíso is located along the slope of steep hills, which encircle the bay towards the north, with a deployment of streets, passages and stairs climbing to the top. At the bottom, on the flat area of the city, are located the business section and the main port of the country.

Coastal section between Tunquen and Algarrobo. This coastline is quite smooth, with sandy beaches. The littoral plains form an undulated landscape of very dissected flattened summits. A low plain extends between Algarrobo and Las Cruces and alternates with large beaches exposed to southwest winds forming extended dune fields.

These beach resorts are characterized by many summer houses and a scarce number of residents. They receive a great floating population on summer weekends due to the proximity to Santiago (108 km). The littoral route runs parallel to the shore, 18 km between Las Cruces and Algarrobo.

Algarrobo enjoys a calm sea where a whole variety of aquatic sports are intensely practised; it is also the seat of the main regattas of the area.

Between Cartagena and San Antonio. The whole coast is heavily populated, being Cartagena, after Viña del Mar and San Antonio, one of the most ancient beach resorts.

The coastal route borders the seaside up to the Maipo river bridge.

This area is the most important center of supply of the whole coast to the south of Valparaíso, with an active business sector and a variety of services.

The coastal plains are very dissected due to the drainage. There is a growing urbanization over fixed dunes, for example, at the interior of San Antonio.

References

Mendez, L. M., 1987 *Paisajes y costumbres recreativas en Chile. Valparaíso en el siglo XIX*. Revista Historia No. 22. 1987. Instituto de Historia, Pontificia Universidad Católica de Chile.
Gran Mapa Caminero de Chile. 1984 Servicio Nacional de Turismo Inupal

Consuelo Castro and M. Inés Valenzuela
Institute of Geography,
Catholic University of Chile,
Santiago,
Chile

8. Recreational uses of Québec coastlines

1. Introduction

Estimated at the scale of 1:1 000 000, Québec benefits from 9000 km of coastlines (Gaudreau et Gauthier, 1981). The coastlines of tourist region 18 (Figure 1), or the region that encompasses all of northern Québec, is not equipped with any recreational installation whatsoever. Docks and small ports represent the only form of shelter available to hardy boaters. On the other hand, the 7000 km of coastlines of the St. Lawrence River system, corresponding to all southern Québec, are relatively well equipped. By the St. Lawrence River system, we refer to the gulf (tourist region 01, 17 and five-sixth of region 02), the maritime and briny water estuaries affected by saltwater tides (tourist region 04, 07 and two-third of region 06), the limnetic estuary affected by fresh water (tourist regions 03, 05, 16, one-sixth of region 02 and one-third of region 06) and part of the river itself up to the province of Ontario border (tourist regions 09, 10, 11 and 12) (Figure 2).

There are no overall studies dealing with land-use along the coastlines of Québec, even more so with regards to recreational activities. However, there are a few general inventories for the southern part of Québec. The most important is a computer-based data bank of Québec's recreational installations (Petitclerc, 1977, 1981). Unfortunately, no historic record of the bank was kept and no systematic update of the data has been undertaken since 1983. The 48 maps at the scale of 1: 50 000 derived from the data bank were probably used for producing the 9 tourist and recreational maps at the cale of 1: 250 000 (Ministère de l'énergie et des ressources du Québec, 1981–83). However, installations belonging to the federal government cannot be found on these documents, as well as the network of parks, reserves, wildlife refuges and resting areas. One must leaf through numerous reports in order to locate these infrastructures. Among the more important of these reports are those produced by: Fisheries and Oceans Canada (1983), Ministère de l'énergie et des ressources du Québec (1983), Ministère de l'environnement du Québec (1983), Fédération de voile du Québec (1984), Ministère du loisir, de la chasse et de la pêche du Québec (1986), Boudreau et Goudreau (1987) and Ministère du tourisme du Québec (1986–87).

The data presented in this paper is thus partial and presents a general outlook of the prevailing situation between 1983 and 1987. The data does not permit to estimate the exact length of the coastlines nor the area affected, but nevertheless gives a rather precise picture of the number and the importance of the installations.

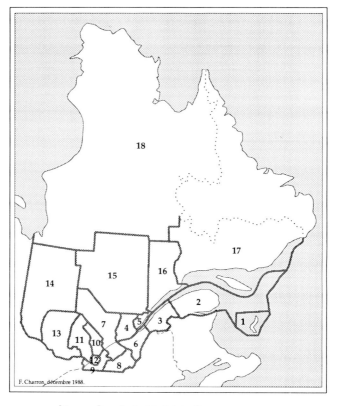

01	– Iles-de-la-Madeleine*
02	– Gaspésie*
03	– Bas-Saint-Laurent*
04	– Québec*
05	– Charlevoix*
06	– Pays-de-l'Érable*
07	– Coeur-du-Québec*
08	– Estrie
09	– Montérégie*
10	– Lanaudière*
11	– Laurentides*
12	– Montréal*
13	– Outaouais
14	– Abitibi/Témiscamingue
15	– Saguenay/Lac Saint-Jean/Chibougamau
16	– Manicouagan*
17	– Duplessis/Ile d'Anticosti*
18	– Nouveau-Québec/Baie James

*   coastal regions specified in this paper

Figure 1. Tourist Regions of Québec.

Table 1. Classification of recreational installations.

1.	Installations associated with other types of recreational installations – Picnic area – Tourist information booth – Athletic field – Public park – Nature study area
2.	Rest area
3.	Golf course
4.	Summer camp area
5.	Camping/trailer park
6.	Public beach

7.	Nautical installations:
	– Boat rentals
	– Boat service area
	– Marina
	– Water access:
	  – Boat ramp
	  – Dock, pier and floating dock
	– Anchorage area

8.	Protected area: – National park – Provincial park – National wildlife reserve – Urban regional park – Migratory bird refuge – Migratory bird resting area – Ecological reserve – Anticosti Island reserve
9.	Summer cottages

Table 2. Distribution of rest areas and facilities by tourist region.

Installations Regions	Sites	Cooking shelters	Benches	Drinking water	Outdoor fireplaces	Tables	Toilets with running water
Gaspésie	16	3	76	4	14	171	21
Bas-Saint-Laurent	2	-	2	-	21	34	2
Pays-de-l'Érable	4	-	-	-	12	47	6
Montérégie	2	1	4	-	2	17	2
Montréal	1	-	-	1	-	40	-
Manicouagan	6	1	12	2	2	38	8
Duplessis	2	-	-	-	-	1	-
Total	33	5	94	7	51	348	39

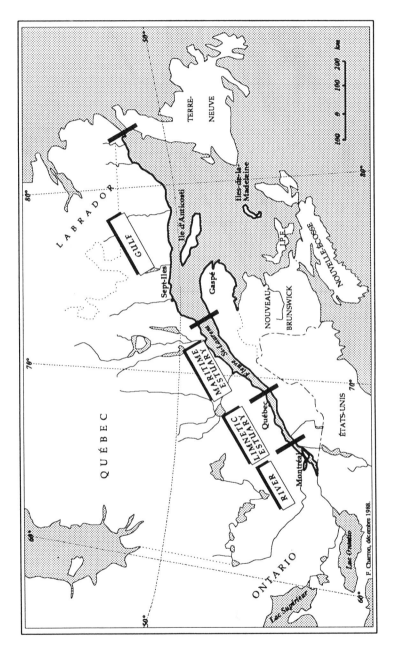

Figure 2. The St. Lawrence River System.

2. Analysis and classification of recreational data

The data was classified according to 9 groups corresponding to the 26 different types of recreational installations most widely found along the coastlines (Table 1). The data was also regrouped according to the tourist regions of Québec as defined by the Québec Provincial Government (Figure 1). We must bear in mind that tourist regions are equipped with many other types of installations which are not associated with coastline activities and that these are generally more numerous. Furthermore, many installations that were too limited in scope or too much associated with the urban environment were not displayed, for example the few historic sites present along the coast or food and lodging facilities. Finally, winter activities were not specifically addressed since these do not require extensive infrastructures. These activities are mainly related to ice-fishing, skating, ice-sailing, snowmobiling and cross-country skiing, etc.

3. Distribution of recreational installations

3.1. Installations associated with other types of recreational installations. We did not discriminate specifically among the following installations: picnic areas, tourist information booths, athletic fields, public parks and wilderness areas. Picnic areas are most often associated with rest areas, campgrounds and parks. Tourist information booths are almost always associated with rest areas or parks. Athletic fields are generally part of campgrounds, summer camps or public parks. Public parks are, for their part, closely linked to the urban environment while wilderness areas are associated with parks in general, or with reserves and refuges.

3.2. Rest areas. Rest areas are more frequently found in the Gaspésie region since the area is one of Québec's oldest tourist regions and traffic is usually more important along the coastal route (Table 2). Most of the rest areas are the responsibility of the Québec government (26 sites), although few facilities are managed by municipalities, the National Park Service or even private enterprise. Most are well designed although few, often too small, constitute eyesores particularly when overfilled with cars and when designers have resorted to shoreline encroachments to gain land area.

3.3. Golf courses. Of the more than 298 golf courses found in Québec, 12 are located in coastal areas; only 3 are associated with campgrounds (Table 3). Golf courses are generally located close to urban centres and thus can be found in only 6 of the 13 coastal regions. Golf courses contribute to the preservation of green spaces in the vicinity of urban centres.

3.4. Summer camp areas. The 14 summer camp areas inventoried can be found in 7 of the 13 coastal regions, not too close to urban centres but not altogether in remote regions (Table 4). Summer camps are established in order to provide a link between man and nature. Installations are thus generally rudimentary. However, many of

Table 3. Distribution of golf courses and facilities by tourist region.

Installations Region	9 hole courses	18 hole courses	Courses associated with campground	Training courses
Gaspésie	2	1	1	2
Québec	1	-	-	-
Pays-de-l'Érable	1	-	-	-
Coeur-du-Québec	1	1	1	-
Montérégie	-	4	1	2
Montréal	-	1	-	-
Total	5	7	3	4

Table 4. Distribution of summer camps by tourist region.

Region	Number of camps	Number of beds per camp
Gaspésie	1	140
Bas-Saint-Laurent	1	120
Québec	1	170
Charlevoix	2	60–150
Coeur-du-Québec	1	80
Montérégie	6	30–700
Manicouagan	2	16–75
Total	14	1415

these areas are equipped with swimming pools or provide access to public beaches, athletic fields, nature study facilities, docks for sailing or canoeing, and one can participate in activities ranging from horseriding to archery.

3.5. Camping/trailer parks. As in the case of rest areas, camping/trailer parks are mainly located in the Gaspésie region where 33 of the 82 areas inventoried are found (Table 5). Most are privately owned and operated (68), although 3 are managed by the government and the remaining are under municipal jurisdiction (11). However, it must be underscored that many of the private sites tend to be neglected. For example, 17 areas in the Gaspésie region offer only essential services (Arsenault, 1974).

The 20 areas with the best installations are located near the larger urban centres. In addition, 21 areas provide access to public beaches and 7 to docks, which are all maintained by their owners. Most of the areas are equipped with toilets, showers, campfire sites, canoeing facilities or a public recreation hall. A large number of these areas are equipped with washers/dryers, a grocery store, a restaurant or a swimming pool.

Table 5. Distribution of camping/trailer parks by tourist region according to the number of campsites.

Region	1–49	50–99	100–149	150–199	200–249	250–299	300+
Gaspésie	15	9	5	3	-	1	-
Bas-Saint-Laurent	2	-	-	-	-	-	-
Québec	2	1	-	-	-	1	-
Charlevoix	-	-	1	-	-	-	-
Pays-de-l'Érable	1	2	2	1	-	1	1
Coeur-du-Québec	2	1	1	-	1	-	-
Montérégie	3	1	3	-	2	-	1
Lanaudière	1	1	-	-	-	-	-
Laurentides	3	-	-	-	-	-	-
Montréal	1	-	-	-	-	-	-
Manicouagan	-	4	-	-	-	-	-
Duplessis	5	-	-	-	1	-	-
Iles-de-la Madeleine	1	2	-	-	-	-	-
Total	36	21	12	4	4	3	2

3.6. Public beaches. Only 32 public beaches were found, but it is evident that this category of recreational installation is under-estimated. It is believed that there are at least ten times more such installations. The 32 areas are those that are considered as rightfully managed: 19 are located in the Gaspésie region, 5 in Manicouagan, 4 in Coeur-du-Québec and 1 in each of the regions represented by Charlevoix, Montérégie, Duplessis and Iles-de-la-Madeleine. Of these 32 areas, 21 are adjacent to campgrounds and 2 are accessible through public parks. Public beaches are mostly sandy beaches or sometimes gravelly beaches, and represent 18.4 km of coastline. Nevertheless, Québec is renowned for having hundreds of kilometers of beautiful sandy coastlines...

3.7. Nautical installations. Such a vast continental and semi-maritime water area as the St. Lawrence River system offers considerable opportunities to boaters. However, since increasing knowledge of navigation is required as one moved downstream, the number of facilities specifically designed for boating decreases accordingly and boaters must depend on public docks, fishing ports and anchorages that lack proper installations to meet their needs (Table 6). For example, we have located 102 access points to the river in the area of the Montreal Urban Community which corresponds to the territory defined by the four following tourist regions: Montégérie, Lanaudière, Laurentides and Montréal. In this area, we find half of all the marinas in Québec, one-fourth of all docks and one-third of all boat-ramps. In contrast, the four remote regions constituted by Gaspésie, Bas-Saint-Laurent, Duplessis and Iles-de-la-Madeleine have 113 access points but only one-fourth of the total number of marinas. Boaters must depend on docks (almost half the number of docks in Québec) and boat-ramps (one-third the number of ramps in Québec).

Table 6. Distribution of docks, marinas and boat ramps by tourist region.

Installations Region	Docks	Marinas	Boat ramps
Iles-de-la-Madeleine	8	3	3
Gaspésie	50	12	27
Bas-Saint-Laurent	11	4	5
Québec	13	5	9
Charlevoix	10	4	6
Pays-de-l'Érable	11	3	8
Coeur-du-Québec	12	6	8
Montérégie	34	27	23
Lanaudière	7	3	5
Laurentides	4	1	3
Montréal	8	18	9
Manicouagan	9	3	5
Duplessis	22	3	11
Total	199	92	122

Docks and marinas are managed as much by municipal (26%), provincial (25%) and federal (16%) governments as by private enterprise (18%) or semi-public (15%) organizations. Municipalities often use these installations to foster their development through the economic benefits or the expansion of tourist activities they promote.

To meet the growing demand in the field of recreational boating, a program for the renewal, expansion or relocation of docks and marinas has been underway since 1984 (Fédération de voile du Québec, 1984) with the financial help of the Government. Old fishing ports have even been converted into marinas. Of the 291 docks and marinas inventoried, 78% of the docks are less than 200 m in length (Table 7). Of the 6 docks extending over 1000 m in length, most are commercial docks.

Table 7. Length (m) of docks according to type of ownership.

Ownership Length (m)	Federal	Provincial	Municipal	Private	Semi-public	Total
1–100	28	34	35	43	16	156
101–200	14	22	22	5	15	78
201–300	3	7	10	1	7	28
301–400	-	2	5	-	2	9
401–500	1	5	-	-	-	6
501–600	-	-	4	-	1	5
601–700	-	-	-	-	-	-
701–800	-	-	-	-	3	3
Others	-	2	1	3	-	6
Total	46	72	77	52	44	291

Table 8. Protected areas in southern Québec.

Type	Number	Area (ha)	Length of coastlines (km)	% of the coasts of St.Lawrence River
National parks	2	33 480	296	4,5
Provincial parks	6	35 450	246	3,8
Urban regional parks	6	726	18	0,3
National wildlife reserves	3	3 308	45	0,7
Ecological reserves	5	1 827	30	0,5
Anticosti Island wildlife reserve	1	794 300	404	6,0
Wildlife refuges	10	25 438	98	1,5
Wildlife resting area	3	654	26	0,4
Total	36	895 183	1 163	17,7

It is evident that recreational boating is a source of pollution since it remains rather easy to empty the contents of the holding tank of a boat once offshore. The fact that pump-out stations are more readily found today does not seem to provide a solution to the problem, since most of the untreated waste-water return to the St. Lawrence eventually (Desjardins-Ledoux and Ledoux, 1977).

3.8. Protected areas. Protected areas cover 18% of the coastlines of southern Québec (Table 8). National and provincial parks, in addition to the Anticosti Island wildlife reserve account for 14% of the total coastlines. Regions with the largest amount of protected areas for conservation are evidently the remote regions of Gaspésie, Manicouagan and Duplessis. However, regions with the greatest number of protected areas for recreation purposes are those located close to Montreal, i.e. Montréal, Montérégie, Lanaudière and Laurentides. In these regions, installations use lesser space because of urban pressures.

In general, parks are the only areas permitting recreational activities outside the limits of their conservation areas. They often provide nature study facilities.

Management of protected areas is the responsibility of the different levels of government, although a few small islands are managed by individuals.

3.9. Summer cottages. The expansion of summer cottages constitutes one of the oldest forms of recreation along the coastlines of southern Québec and relates basically to beach and boating activities. During the 19th century, members of the English nobility were almost the only users of this form of recreation which was rather expensive for the times. Now, the cost of owning a cottage is within reach of many individuals and most of the sites presenting good potential for recreation, specially sandy beaches, are already built-up (Arsenault, 1974).

In 1979, Saint-Amour estimated that there were 170 000 cottages in Québec (Table 9). Unfortunately, we do not have data relating to individual tourist regions,

Table 9. Distribution of summer cottages by administrative region.

	Administrative region	Cottages	%
1-	Bas St-Laurent-Gaspésie	4 001	2,39
2-	Saguenay-Lac St-Jean	8 463	5,05
3-	Québec	22 838	13,63
4-	Trois-Rivières	12 982	7,71
5-	Cantons de l'Est	10 469	6,25
6N-	Montréal (nord)	60 484	36,10
6C-	Montréal (centre)	30 002	1,79
6S-	Montréal (sud)	14 220	8,49
7-	Outaouais	27 477	16,40
8-	Abitibi-Témiscamingue	1 630	0,97
9-	Côte-Nord	2 002	1,19
	Québec	167 568	100,00

Saint-Amour (1979, p. 57).

nor to the specific areas of the coastlines of the St. Lawrence River system which are affected. It is however certain that the majority of these cottages are located along the shores of the St. Lawrence or besides rivers and lakes.

Summer cottages rarely enhance the natural environment. Vegetation is often replaced by structures, shoreline deforestation results in erosion, encroachments are numerous, rubbish and waste waters are disposed indiscriminately, and installations do not always provide a pleasant sight.

Conclusion

The distribution of recreational installations on the coastlines of Québec is irregular since it varies according to the location of the centres of demand, the needs of the population and the climate. As we have seen, it is rather difficult to assess the impact of recreational installations in quantitative and even qualitative terms. We can only derive an approximation as to the total area of the coastlines affected by recreational installations, a value which corresponds to close to 30% of the total area. As to the quality of the installations and sites, all remains to be investigated.

5. References

Arsenault, G. (1974) Land use planning for outdoor recreation in the littoral zone of the St. Lawrence, Mémoire de M.Sc., Département de géographie, Université Laval, 142.
Boudreau, F. and Goudreau, L. (1987) Les milieux naturels protégés au Québec, Ministère de l'environnement du Québec, 28.
Desjardins-Ledoux, R. and Ledoux, A. (1977) La navigation de plaisance au Québec, Éditions de l'Homme, Montréal, 233.
Fédération de voile du Québec (1984) Guide nautique du Saint-Laurent, Montréal, 208.
Ministère de l'énergie et des ressources du Québec (1981–83) Cartes de tourisme et de plein air. 9 maps at the scale of 1: 250 000.

Ministère de l'énergie et des ressources du Québec (1985) Les territoires récréatifs et protégés au Québec. Map at the scale of 1: 250 000.
Ministère de l'environnement du Québec (1983) Les réserves écologiques au Québec, Direction des réserves écologiques et des sites naturels, 37.
Ministère du loisir, de la chasse et de la pêche du Québec (1986) Analyse de la répartition des équipements de loisirs au Québec, Conférence nationale du loisir, 27.
Ministère du tourisme du Québec (1986–87) Guides touristiques régionaux.
Pêches et Océans Canada (1983) Guide des ports pour petits bateaux. La Société de cartographie du Québec. Map at the scale of 1: 250 000.
Petitclerc, R. (1977) Essai concernant la définition et la symbolisation des termes touristiques utilisés dans S.I.R.T.E.L., Ministère du tourisme, de la chasse et de la pêche du Québec, 69.
Petitclerc, R. (1981) S.I.R.T.E.L. (Système d'Inventaire des Ressources Touristiques et des Équipements de Loisirs et de plein air). Ministère du loisir, de la chasse et de la pêche du Québec. 48 maps at the scale of 1: 50 000.

Jean-Marie M. Dubois and Marc Chênevert
Département de géographie et télédétection,
Université de Sherbrooke,
Sherbrooke,
Québec,
Canada J1K 2R1

SECTION II

Coastal Recreation in Adverse Environments

9. Recreational use of the Washington State coast

Introduction

Though it is only 300 km from Washington State's southern to northern borders (from the Columbia River to Canada respectively), the marine coast of the State is 4,334 km in length. Combined with a temperate northwest-coast climate, this extensive coastline provides a recreational paradise for the 4.2 million people who live in the state.

The area can be subdivided into three regions (Figure 1): the Pacific Coast, the Strait of Juan de Fuca coast, and Puget Sound with its adjoining Hood Canal fiord. While the geomorphology of these sectors has been reported upon in detail (Downing, 1983; Schwartz and Terich, 1985; Schwartz et al., 1985; Terich and Schwartz, 1981), this report will deal primarily with their recreational use. In this context, it is appropriate to consider such separate uses as boating, parks and reserves, beaches, and tourism.

Boating

In terms of sheer numbers, population and investment, boating probably leads all recreational uses. It is estimated that one out of every three people in the state participate in boating activities, with either their own or rented boats. The craft involved vary from small rowboats and kayaks to large power and sailing vessels. Generally speaking, motorized and sailing craft over 5 m in length are subject to registration and excise tax fees. Washington is in compliance with the Federal Boat Safety Act and, in 1984, qualified for approximately $250 000 in government boat safety aid.

Available data on marine boating amenities in Washington are listed as follows: marine boating parks (Table 1), and marina-moorage-storage facilities (Table 2).

Parks and reserves

Government agencies, at different levels, maintain a great many coastal recreation parks in the State of Washington. These offer, in varying degrees, picnic, parking and game-playing areas, restrooms, campsites, nature trails, interpretive signs and centers, boat launches and beach access. The Washington Marine Atlas series, issued by the Department of Natural Resources, lists 428 shoreside public parks and public access beaches and tidelands. No resident of western Washington is ever far from a place to stand at the shore.

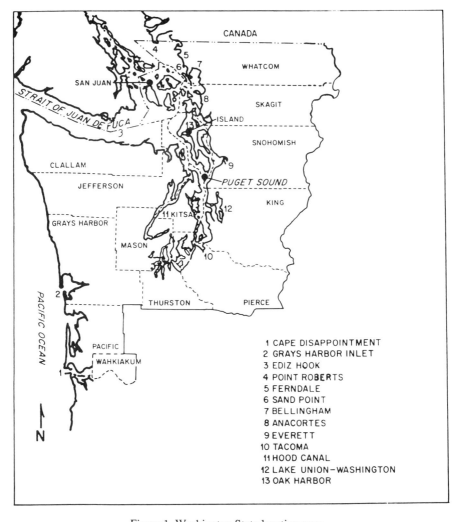

Figure 1. Washington State location map.

For *Scuba* diving, the Washington State Parks Department has established underwater parks at:

1. Peapod Rocks, San Juan Islands
2. Deception Pass
3. Orchard Rocks, Ft. Ward
4. Blake Island
5. Saltwater State Park
6. Tolmie Beach State Park
7. Kopachuck State Park
8. Ft. Casey State Park

Table 1. Marine Parks guide

Puget Sound Marine Parks:	Mooring buoys	Float space (footage)
Blake Island[1,2]	21	1562
Cutts Island[1]	9	–
Eagle Island[1]	3	–
Fay Bainbridge[3]	2	–
Fort Flagler (Kilisut Harbor)[3]	7	256
Fort Ward[2,3]	2	–
Fort Worden[2,3]	9	128
Illahee (Sandy Beach)	5	415
Jarrell Cove	14	642
Kitsap Memorial	2	–
Popachuck[2]	3	–
McMicken Island[1]	5	–
Mystery Bay	7	628
Old Fort Townsend	4	–
Penrose Point	8	225
Pleasant Harbor	–	220
Potlatch (south side)	5	–
Sequim	6	424
Squaxin Island[1]	10	340
Stretch Point	5	–
Tolmie[2]	5	–
Twanoh (west side)[3]	5	84
San Juan Marine Parks:		
Blind Island[1]	4	–
Bowman Bay	5	102
Clark Island[1]	9	–
Cornet Bay	0	645
Doe Island[1]	–	60
Hope Island	4	–
James Island[1]	5	90
Jones Island[1]	7	260
Matia Island	2	128
Patos Island[1]	2	–
Posey Island[1]	–	–
Saddlebag Island[1]	–	–
Skagit Island	2	–
Spencer Spit	16	–
Stuart Island[1]		
(Reid Harbor)	15	219
(Prevost Harbor)	7	539
Sucia Island[1]	(Total)	692
(Echo Bay)	14	
(Fox Cove)	4	
(Shallow Bay)	8	
(Ewing Cove)	4	
(Snoring Bay)	2	
(Fossil Bay)	16	
Turn Island[1]	3	

[1] accessible only by boat.
[2] includes underwater park for scuba divers.
[3] boat launch.

Table 2. Marine facilities in Washington State in 1981 (after Goodwin, 1982).

County Region	Total marinas and dry storage facilities			Total wet moorage slips		
	Public	Private	Total	Public	Private	Total
Whatcom	2	12	14	1 166	1 464	2 630
San Juan	3	18	21	123	841	964
Skagit	2	13	16	853	1 325	2 178
Island	3	10	14	316	173	489
Snohomish	2	13	15	2 942	93	3 035
King*	5	74	79	3 141	4 775	7 916
Pierce	0	36	36	0	3 300	3 300
Mason	5	9	14	96	109	205
Thurston	1	11	12	0	1 433	1 433
West Clallam	1	18	19	344	510	854
Pacific and Grays Harbor	7	9	16	1 750	131	1 881
East Clallam and Jefferson	2	10	12	932	849	1 781
Kitsap	8	21	29	1 069	940	2 009
Wahkiakum**	5	3	8	944	70	1 014
Total	46	258	304	13 676	16 013	29 689

* Includes Lake Union and Lake Washington.
** Includes portions of two adjoining counties on Columbia River estuary.

There are also numerous other recommended underwater recreation areas open to the public.

A difficult to quantify recreational use, that will be discussed briefly here, is sport fishing and the collection of shellfish. The fishing, usually connected with recreational boating, is concentrated mostly upon the abundant salmon runs. Shellfish collecting (oyster, clams, mussels, scallops) is carried out, mainly at low tides, on the many thousands of hectares of tidal flats and lower beach foreshores in the three regions delineated in the introduction to this report. Trapping the very plump and famous Dungeness crab also adds another dimension to the collection activity. Vast amounts of seafood are collected and consumed in this manner, and its importance as a form of recreation cannot be overlooked.

National Wildlife Refuges (N.W.R.), large tracts of land set aside by the federal government for protection of wildlife and their natural habitats, comprise a major component of the recreational coastal areas in Washington State. The main uses by people include birdwatching, hiking, photography, nature trails, and the scenic vistas; with, to a lesser extent where permitted, swimming, boating and fishing. In Washington there are:

	Site	Region
1.	Dungeness N.W.R.	Strait of Juan de Fuca
2.	Flattery Rocks N.W.R.	Pacific coast
3.	Quillayute N.W.R.	Pacific coast

4.	Needles N.W.R.	Pacific coast
5.	Copalis N.W.R.	Pacific coast
6.	Nisqually N.W.R.	Puget Sound
7.	San Juan Islands N.W.R.	Puget Sound
8.	Willapa N.W.R.	Pacific coast

Beaches

As stated in the previous section, the Washington State Department of Natural Resources has identified 428 marine park sites. In addition, there are innumerable beach access sites, identified by colorful signs, that have been designated in a major undertaking by the State Department of Ecology. These beach locations range from a few meters of public foreshore in some corner of Puget Sound to the very long stretches of public beach on the Pacific coast.

Due to the chilling influence of the Japanese (Kurukio) Current, and the rather northerly latitude, water in Puget Sound and off the Pacific coast is quite cold during the summer months. Though swimming is indulged in only by the hardy, people throng to the beaches throughout the year. There is always beachcombing or walking, the beautiful scenery, auto-cruising on some of the Pacific beaches; and seasonably, shellfish gathering. Then, too, the northern sector of the Pacific coast, with its many sea cliffs, stacks, and arches, is one of the most scenic coasts in the world.

Beach attendance in 1984, compiled by the Washington State Parks and Recreation Commission, indicates considerable traffic at a few selected sited.

	Site	Attendance
1.	South Beach	2 840 206
2.	Deception Pass	2 392 411
3.	Long Beach	1 891 506
4.	North Beach	1 618 078
5.	Birch Bay	1 280 738
6.	Mukilteo	1 221 002

When considering the regional population, this is a heavy concentration, indeed.

For a comparison of the coastline length of the various marine counties in western Washington, see Table 3.

Tourism

Regional, national and foreign visitors to the coast of Washington come by land, sea, and air. What they seek in the way of coastal recreation would include the aforementioned boating, parks, reserves, and beaches; but, would also cover convention meetings, sightseeing, seafood restaurants, local celebration events, picturesque old homes and towns, and the considerable wine industry that has been nurtured by the mild marine climate.

Table 3. Coastline length of the marine counties in western Washington (Washington State Dept. of Natural Resources).

County:	Length (km):
Clallam	355
Grays Harbor	210
Island	354
Jefferson	140
King	181
Kitsap	394
Mason	349
Pacific	320
Pierce	371
San Juan	602
Skagit	291
Snohomish	130
Thurston	178
Whatcom	189

While it is difficult to separate coastal tourism from that in the rest of the state, it is safe to assume that since most of the population and tourist facilities are located in the marine counties, then most of the tourism occurs there as well.

The following figures for 1984, compiled by the Tourism Development Division of the U.S. Travel Data Center, give us a breakdown on tourism in Washington State.

Expenditures	$ 3 billion
Payroll	$630 million
State Tax Receipts	$128 million
State Budget	$2.3 million
Visitors:	
out-of-state	15 million
in-state	10 million

Where the foreign tourists came from can be ranked as follows:

Rank	Country	Number
1.	Canada	1 659 000
2.	Japan	60 000
3.	United Kingdom	30 000
4.	Taiwan	22 000
5.	Hong Kong	20 000
6.	Germany	14 600
7.	Sweden	10 000
8.	Norway	7 500
9.	Australia	5 000
10.	France	4 000

Obviously, international tourism plays an important role in the State of Washington. The economic impact is considerable, and can be tabulated as follows:

Total foreign arrivals 1 830 000
(4th highest state)
(9% U.S. total)

Total expenditures $217 million
(11th highest state)
(7% U.S. total)

Total Payroll $ 50 million
(12th highest state)

Total employment 5 600
(11th highest state)

State tax revenue $ 12 million

Table 4. Tourism impact upon marine counties of Washington State during 1984 (U.S. Travel Data Center).

County	Expenditures ($millions)	Payroll ($million)	Employment	State tax ($million)	Local tax ($thousand)
Clallam	36	6.3	860	2.1	310
Grays Harbor	40	8.0	1 100	1.9	380
Island	10	1.5	300	.2	40
Jefferson	11	2.0	300	.6	100
King	1 450	310.0	35 000	64.0	11 000
Kitsap	31	7.7	920	2.1	310
Mason	13.8	3.1	365	.73	210
Pacific	10	1.9	285	.59	99
Pierce	500	100.0	5 500	8.8	3 300
San Juan	30	7.0	700	2.0	300
Skagit	23	4.0	600	1.0	100
Snohomish	77	14.0	2 100	3.6	820
Thurston	52	10.0	1 560	3.1	520
Whatcom	37	7.0	1 200	2.1	310

For a more detailed, county-by-county, survey of tourism in the marine counties of western Washington, see Table 4.

References

Downing, J., 1983, The coast of Puget Sound: Washington Sea Grant Program, Seattle, 126.
Goodwin, R., 1982, Recreational boating in Washington's coastal zone; the market for moorage: Washington Sea Grant Marine Advisory Program, Seattle, 134.
Schwartz, M. L., Mahala, J., and Bronson, H., 1985, Net shore-drift along the Pacific coast of Washington State: *Shore and Beach* **53**, 21–25.
Schwartz, M. L. and Terich, T., 1985, Washington, in Bird, E. C. F., and Schwartz, M. L., (eds.), *The World's Coastline*: Van Nostrand Reinhold, New York, 17–22.
Terich, T. and Schwartz, M. L., 1981, A geomorphic classification of Washington State's Pacific Coast: *Shore and Beach* **49**, 21–27.

Maurice L. Schwartz
Department of Geology,
Western Washington University,
U.S.A.

10. Pacific coast recreational patterns and activities in Canada

Introduction

The Pacific coast of Canada is only some 850 km in length as the crow flies, but the actual length of coast is in the order of 41 000 km taking into account the innumerable islands and penetrating coastal fiords. Recreational use of this resource is limited by three main factors: topography, climate and accessibility. Topography is steep with deeply incised coastal fiords surrounded by rock walls over 2000 m in height dominating much of the coast. The east coast of Vancouver Island and the Fraser delta provide the only major areas of respite from this rugged topography (Figure 1).

The climate is dominated by the onshore flow of Pacific airstreams which release considerable precipitation in their passage over the mountain ranges. Winters are generally mild and very wet, whereas summers are cool, but less humid. The main exception to this pattern is the recreationally-significant climate of the southeastern coastal portion of Vancouver Island and the Gulf Islands. Here, in the rain shadow of the mountains to the west, a Mediterranean-type climate prevails, giving prolonged periods of warm, sunny weather, ideal for boating and other recreational activities in the summer period.

Access to the coast, other than by boat, is limited. Major populations are located in the southern part of the area around Vancouver and Victoria, with larger populations to the south again in the U.S. centred around Seattle. By far the majority of recreational activities take place in close proximity to these centres whether they are water or land-based marine activities. Most of the coast north of Vancouver Island is inaccessible from the land. Settlements are few and far between. Recreationists must be self-contained. Wilderness type activities and participants dominate. The area probably constitutes one of the finest in the world for marine-based wilderness recreation activities.

This paper will describe the recreational activities and patterns occurring on this Pacific coast of Canada and summarise the major problems. The description will use the framework shown in Figure 2 separating the activities into land-based and water-based.

Water based activities

The coastal waters of British Columbia are not warm, meaning that water-based activities tend to be located on rather than in the water. Activities can be classified according to this distinction, with by far the majority of recreational use being on the water in the form of boating, and a much smaller amount being actually in the

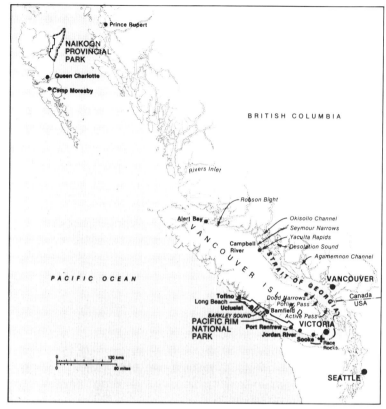

Figure 1. The Pacific coast of Canada.

water, usually with the participants wearing some sort of thermal protective wear, such as in scuba diving and surfing. Exceptions do occur, however, with free swimming being popular in sheltered and warmer waters in Georgia Strait (Figure 1).

Cruises

Two main kinds of cruises can be identified, depending upon the size of the vessel and goals of the cruise. At one end of the continuum large luxury liners ply the British Columbia coast en route to Alaska. Although the major attractions are scenery and wildlife (Montgomery, 1981), there is little physical contact with the environment and the life on board the vessel constitutes a significant part of the experience. At the other end of the spectrum, numerous smaller vessels provide a limited, but equally high-paying, number of passengers the opportunity for extensive physical contact with the environment. Education is a major goal for such cruises. Both market segments have experienced very rapid growth. A paucity of data makes this difficult to measure for the small operators, although the growth in

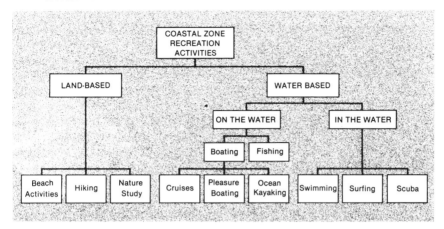

Figure 2. Organisational framework for coastal zone recreation activities on the Pacific coast.

advertising for such cruises provides indirect evidence, whereas the luxury 'Love Boat' cruises increased at an annual rate of 60% between 1970 and 1981 (Montgomery, 1981).

'Love Boat cruises'

The Inside Passage cruise to Alaska is one of the most lucrative in the world. The piracy of boats and planes and the hijacking of the Italian cruise ship *Achille Lauro* has done much to boost cruise boat traffic on the route, mainly at the expense of the Mediterranean competition. Princess Cruises of Los Angeles for example, dropped 18 Mediterranean cruises in 1986 with a potential revenue of $43 million and moved its most famous boat, the *Pacific Princess*, featured on the television series *The Love Boat*, to Seattle, Washington. In 1983, 16 vessels were engaged on the Inside Passage route, by 1986 this had risen to 21 carrying over 325 000 passengers. It is estimated that they contributed over $130 million to the economy of Vancouver, (Daniels, 1986). No data is available on the economic impact on other smaller ports although it is estimated that Victoria, the provincial capital, generated $3 million in 1983 (Dufour, 1985). The effects on other smaller communities, such as the native settlement of Alert Bay (Figure 1), will be much less in total but probably much more significant in relative terms to these isolated and relatively poor communities.

In terms of its purely recreational impact the luxury liner trade is not significant. Although the passengers are undertaking a recreational experience for which they are willing to pay large amounts of money, the principal recreational environment is the actual vessel, a strictly controlled and artificial environment. However, it should not be forgotten that the coastal scenery is the factor that has made the Alaska run one of the most popular in the world. Furthermore, the 1980 ruling limiting the number of vessels entering Glacier Bay, Alaska, in order to protect whale breeding grounds, serves to remind of the potential impact of the trade on coastal wildlife habitats.

Nature cruises

One of the most rapidly growing aspects of commercial marine recreation is the emergence of a significant nature cruise sector. Operating generally in quite small boats (under 20 m) such cruises immerse the participant in the natural and cultural milieux of the coast. They range widely in scope and cost. Some involve a 3 hour whale-watching trip to see the gray whales on the west coast of Vancouver Island during their annual migration, others focus on sea lions or killer whales. Longer cruises may be up to 3 weeks in length and encompass a much broader range of enquiry.

These trips are part of the more general recreational phenomena of recreational education that has also benefited membership in conservation organizations and the sale of nature guide books of all descriptions, binoculars, cameras, and other equipment. They cater to the educated, middle classes interested in learning more about their environment, as a recreational experience. One operation, Pacific Synergies, advertise:

> If you're the sort of person for whom the very idea of an educational holiday sounds like a contradiction in terms, look into our sail expeditions in the Queen Charlotte Islands. A good many people have had the time of their lives living and travelling aboard our 45-foot sailing cutter while learning about the people, history and archeology of the little-known and largely unspoiled Queen Charlottes, the ancestral home of the Haida people.

The operators of this boat purchased a larger 22 metre boat in 1985 due to extra demand and have for their 1988 cruises also added an additional vessel, a 30 metre schooner. Over the last six years they have had over 1000 passengers on their trips around South Moresby Island in the Queen Charlottes. The boat has an extensive library, diving equipment and microscopes and usually sails with an on-board naturalist. Average trip cost for a 10 day trip is $1750 and demand far exceeds supply for berths. Survey research by the author (Dearden, 1987a) indicates that most participants are motivated by the wilderness qualities of the area and the mode of travel, by sail. The trips have been described in outlets such as *National Geographic* and the *New York Times*.

These trips are not unique. Other operators now run similar trips, with the Queen Charlotte Islands being a particularly popular location. This traffic will increase considerably in the years to come as the South Moresby portion of the Islands is now under negotiation to establish a new national park reserve. This area, known as the 'Canadian Galapagos Islands' due to the high number of endemic species, received large amounts of publicity on the national and international scenes due to the highly emotional conflicts between native Indian interests, environmentalists and the forestry industry over whether the area should be logged or made into a park (e.g. see Dearden, 1987b). There are no roads or trails. The only access into this archipelago of some 138 islands is by boat.

Another related kind of experience, but of shorter duration, are the increasing numbers of trips organised for viewing marine mammals, particularly sea lions (e.g. Race Rocks, Tofino), gray whales (Ucuelet, Tofino) and killer whales (Robson Bight, San Juan Island). Canada's Pacific coast is one of the best areas in the world to view killer whales, particularly at the Robson Bight area between Vancouver Island and the mainland (Figure 1). Here viewers can be virtually guaranteed to encounter the whales everyday between June and early September. The area is not accessible by road and visitors from all over the world come to see the whales either with their own boats or to book a place on a charter boat. Over 1000 such visitors came to see the whales in 1986 with an average gross trip expenditure of $660 per person (Duffus, 1988). A charter berth cost $50. Similar trips are also run on the west coast of the Island to view the gray whales during their spring migration (March/April) and also during summer residency. Such trips are less expensive ($25 per trip) and are usually composed of recreationists for whom whalewatching is just one part of their trip to the west coast. In excess of 6000 people take such trips each year with an average gross trip expenditure of $175 per person (Duffus, 1988).

Pleasure boating

There are more boats per capita in British Columbia than in any other province in Canada, and 70% of the boat owners are found within the Strait of Georgia region (Harrison, 1979). In addition to the self-owned, resident category, a further five percent of households rent boats and substantial numbers of boaters from the United States visit British Columbia waters. Accurate data on the latter are not available, but Cooke's research on transient boat populations cruising in Georgia Strait revealed the numerical importance of non-resident boaters and the significant increases in their numbers that can be expected (Cooke, 1979). A survey by Rhodes (1979) of non-resident boating populations on the attraction of British Columbia waters listed 21 factors mentioned by respondents, with scenery, sheltered waters, fishing, less crowding, and birds and wildlife being the top five features. Figure 3 shows some trends in boating populations.

This large number of boaters, both resident and non-resident, requires various facilities, such as moorage, fuel and supplies. These are provided both by private enterprise in the form of marinas, launching ramps, fuel barges, stores, restaurants and similar facilities, and by government in the provision of moorage facilities and launching ramps.

Two main kinds of moorage facilities can be recognized: permanent, where the boat is kept, and transient, for cruising purposes. Substantial deficits in the provision of both types of facility may be experienced in the future. It was calculated in 1978 that fixed costs to resident boat-owners, such as moorage, insurance, and repairs, totalled some $61 million, and that operating costs, such as fuel, food and temporary tie-up fees, totalled $134.3 million. Non residents contributed a further $12 million in operating costs (Canada and British Columbia, 1979). These figures will have grown considerably since this time. Obviously the provision of adequate onshore activities is critical in maintaining this high economic return.

Private marinas, of which there are over 200 along the coast, are located mainly in the lower Georgia Strait area. They vary greatly in size from fully serviced docks with attached resorts and restaurants to small floats and launching ramps. Cooke

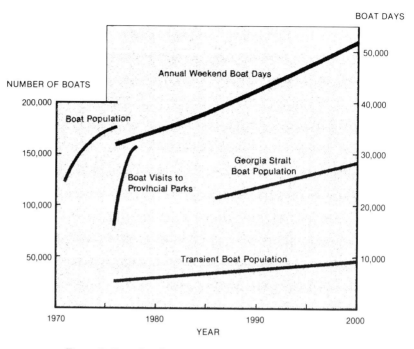

Figure 3. Some boating trends on Canada's Pacific coast.

(1977) found that 63% of the transient boaters interviewed preferred such private facilities compared with public ones, despite the higher cost. Indeed, the most isolated facilities provided by the government rarely require fees of any kind for overnight moorage. Such wharves, administered by the Small Craft Harbours Branch, also vary greatly in degree of service provided, but with less likelihood of having on-dock power and water than the private facilities. This difference, perhaps, accounts for the preference for private facilities noted above. The Department of Transport also provides simple floats with no services, or charges, in strategic locations as ports of refuge in inclement weather conditions. Although these federally provided facilities can be used by pleasure craft, they are principally designed for small commercial boats and, in particular, fishermen.

By way of contrast, the provincial government provides facilities largely for the recreational boater. Primary among these are the marine parks scattered mainly along the lower coast. This system of provincial marine parks has developed largely on an *ad hoc* basis prompted by boaters' demands, rather than as a preconceived and planned marine park system. The primary criteria for establishment are good anchorage combined with available uplands. Most are within 10 to 20 ha in size and contain little or no facilities, other than perhaps a small dinghy-mooring float. Desolation Sound at the northern tip of the Strait of Georgia is the largest park, with more than 60 km of shoreline, several offshore islands, an upland area containing lakes and waterfalls, and the whole surrounded by snowcapped peaks some 2400 metres in height (see Dearden, 1986 for a detailed discussion of this park and its management problems). Only five of the parks have any attendance records for moorage. However, most of the parks are extremely popular during the summer months and have a growing clientele during the rest of the year.

The federal government is also proposing to create a system of national marine parks. Unlike the recreationally-oriented provincial parks, the national marine parks will be established 'to protect and conserve representative examples of the diversity of Canada's coastal zone and oceans for the benefit of present and future generations', and not be primarily recreation-oriented, (Parks Canada, 1983). Little mention is made of recreational boating in the policy for the parks and it is not anticipated that the creation of national marine parks in British Columbia would influence pleasure boating to the same degree as the provincial parks. Dearden (1987c) has discussed the main forms of marine protective designation available in Canada and their purposes.

The majority of boating activity takes place in the sheltered waters of the Strait of Georgia, with Victoria, Vancouver and the United States being major points of origin and the principal destinations being the Gulf Islands, Sunshine Coast and Desolation Sound. Hence, the major points of origin and destination are in relatively close proximity. Much greater time, effort and experience is required to navigate to more distant cruising grounds along the west coast of Vancouver Island or the mainland coast to the north. The weather, also, is likely to be less favourable in such locations where there are fewer support facilities available. However, these areas are attracting increased numbers of boaters who are perhaps disenchanted with the overcrowded situations in the more popular inland waters.

The final trend in recreational boating to which attention should be drawn is the very rapid growth of ocean kayaking. Empirical data to substantiate this trend are difficult to obtain, but indirect evidence, such as the growth in specialty shops and clubs, public lectures on the topic, articles in boating magazines and personal observation over the last five years, indicates substantial growth. Ocean kayakers were very rarely encountered in the main inland cruising grounds, such as the Gulf Islands, Desolation Sound and Johnstone Strait, five years ago, but now provide a very common sight, particularly in Desolation Sound, Robson Bight, the Broken Group Islands and in the Queen Charlotte Islands. It appears that this form of low impact boating is going to become quite an important mode of coastal recreation enjoyment in the years to come.

Fishing

Estimates suggest that 60% of pleasure boats on the west coast are also used for sport fishing, dominantly for Pacific salmon (Harrison, 1979). Until recently, little attention has been directed to the sport fishery, either as an important component of the salmon resource allocation, or as a major form of coastal recreation. However, recent studies have indicated the importance of the activity especially in economic terms where gross expenditures were estimated at $166 million in 1986 (Department of Fisheries and Oceans, 1987).

The sport fishery is largely based on three of the five species of Pacific salmon – chinook, coho and pink. However, increasingly anglers are beginning to realize the value of other fish, such as snapper and ling cod, both as a fishing challenge and culinary delight. Tidewater fishermen are now required to purchase licences to fish and this has helped provide better data on the numbers participating. For example, 282 247 licences were sold for a total value of 1 730 540 dollars in 1981. This figure had risen to 371 199 by 1987 (Department of Fisheries and Oceans, 1987). It is estimated that the sport fishery takes something like four percent of the total salmon catch and, yet, the sport fishing fleet is worth considerably more than the commercial fleet. This has led many economists to recommend a far greater share of the catch be allotted to the sport fishermen, at the expense of the commercial fishermen (Pearse, 1982).

Geographically most of the sport fishing activity is concentrated within the Strait of Georgia, particularly at the northern end near Campbell River, and also in Juan de Fuca Strait off Sooke. However, salmon fishermen and fishing resorts can be found throughout the area between these two points. North of Campbell River the fishing is generally good, but the remoteness deters many family fishermen. An additional large concentration of fishing activity occurs at Rivers Inlet where anglers come from all over the world to fish for trophy-sized chinooks. A similar fishery is now beginning to develop on the west coast of Vancouver Island in the Barkley Sound area, as a result of the abundant success of enhancement activities in the area.

In-the-water activities

Three main in-the-water activities take place in British Columbia waters. Two of them, surfing and swimming, are somewhat localized and likely to remain that way, whereas the third, scuba diving, appears to be undergoing a major expansion all along the coast. Surfing is an important but rather localized activity occurring mainly on the beaches of the west coast of Vancouver Island, chiefly at Jordan River and Long Beach. By way of contrast, ocean swimming occurs mainly in the warmer waters of the Strait of Georgia. Although tolerable conditions for the hardy exist throughout the Strait in summer, and swimming is a popular activity off many beaches, the waters at the northern end are noticeably warmer.

The third activity, scuba diving, takes place all along the coast although diver numbers are higher in the south. There are some 41 000 active and participating divers in British Columbia as well as an estimated 4600 diving tourists per year, of which 64% are from out-of-province (Ernst and Whinney, 1980). It is estimated that judicious marketing could boost the latter figure by 800 percent to make a considerable increase on the $1 million currently generated by diving-related accommodation and charter boat sales. Furthermore, in terms of economic importance, it should be noted that diving equipment is not cheap, with an average basic kit costing in the region of $1500. In response, there are a large number of diving shops catering to the needs of divers and forming the focal point for diving activity.

The activity is based upon the excellence of the British Columbia marine environment, with a wide assortment of marine life nurtured by the rich nutrients, and miles of rocky headlands and reefs off the coast. Tidal passages provide some of the most spectacular sites for marine invertebrates and areas such as Race Rocks near Victoria, Active Pass, Porlier Pass and Dodds Channel in the Gulf Islands, Agamemnon Channel on the Sunshine Coast, and Seymour Narrows, Yuculta Rapids and Okisollo Channel farther north are becoming known internationally as spectacular diving sites (Figure 1). In addition, the coast has 16 different marine mammals, over 200 species of fish, the world's largest octopus and the greatest abundance of sea star species found anywhere in the world. Despite such abundance, local area closures, harvest limits and seasonal restrictions have to be enforced to protect the wildlife from overzealous collectors. Even so, many populations, such as abalone, crab and ling cod have become severely depleted through over-harvesting in some areas. This is seen to be a major problem in the future if recreational diving is to continue to be a major tourist activity, and emphasizes the need for some form of protective status for the most commonly used diving sites.

Land-based recreational activities

Most land-based recreational use of the coastal zone is concentrated around the heavily populated Georgia Strait region and the southwest coast of Vancouver Island. In these areas extensive use is made of beaches and coastal trails for activities, such as beachcombing, nature study and sunbathing. The east coast of Vancouver Island, between Cowichan Bay and Campbell River, in particular, is intensively developed by beach-oriented resorts, whereas the west coast of the Island provides more opportunity for wilderness-type recreational activities.

Pacific Rim National Park is a major focus for land-based, non-urban, coastal zone recreation in the province. Located on the west coast of Vancouver Island, it is comprised of three geographically separated areas spread along 100 km of the coast (Figure 1). The most northerly unit, and that which receives the bulk of the half a million visitors per year is the Long Beach unit. Most of the park naturalist and interpretation facilities are concentrated in this unit. Broad expanses of beaches make it a favourite for family recreation during the summer, when available campsites within the park are in very short supply.

Off-season camping is also becoming increasingly popular in the area. Due to the relatively mild winter conditions, campers in tents can be found in the campgrounds all year. The total number of visitors is also bolstered by those staying in the comfortable private resorts surrounding the park. Many of these off-season visitors are engaged in various nature-watching exercises. In particular, whale-watching has blossomed over the past few years, with the twice annual migration offshore of the Pacific grey whale (*Eschrictius glaucus*) being a focal point. Up to 50 whales also appear to reside along the Vancouver Island-Washington coasts during summer, rather than making the trek from Mexico to the Arctic and back again, and can sometimes be seen feeding in the shallow waters off the park.

The second unit of Pacific Rim National Park is the Broken Group Islands to the south of the Long Beach unit. It consists of over 100 small islands at the mouth of

Barkley sound. The area is becoming very popular with ocean kayakers attracted by the relative isolation and the contrast between the high energy outer coasts and the secluded beaches, as well as the abundant wildlife of the inner coasts. Although camping there is designated as primitive, there are increasing signs that significant ecological changes are taking place, including the large-scale destruction of vegetation, due to the sheer numbers of people now visiting the islands. The park's administration is also receiving increasing numbers of letters regarding the conflicts between the rising number of users and environmental preservation, and between different kinds of users. These will intensify in the future, as all trends point to increasing numbers of visitors.

These problems of environmental degradation and inter-use conflict are, perhaps, most evident in the third and most southerly unit of the park, the West Coast Trail, a linear section stretching from Bamfield in the north to Port Renfrew in the south. The roughly 80 kilometre long trail is based mainly on the 'lifesaving trail' that was originally built along this 'Graveyard of the Pacific' to help shipwrecked sailors. Since its reconstruction by Parks Canada, the trail has become extremely popular as a coastal rain forest wilderness hike. It now attracts an estimated 8000 visitors per year from all over the world. A recent letter to the Minister of Environment from a hiker who had hiked similar long-distance trails in New Zealand reads in part:

> The West Coast Trail has the potential to be one of the finest wilderness tracks in the world ... Despite its rugged beauty and ocean wildlife ... my enthusiasm for British Columbia's West Coast Trail has diminished ... The litter on the West Coast Trail is, frankly, an embarrassment. At each and every campsite human excrement and toilet paper is found. Paper, plastic, cans, and broken glass despoil the otherwise beautiful environment. The problem is most acute on the hiking trail itself (Almack, 1983).

This is not an isolated viewpoint and indicates one of the severe management problems that will have to be met by Parks Canada in the future. It is also symptomatic of the problems of land-based recreation on the British Columbia coast. Vast areas are inaccessible and sustain virtually no land-based recreational use, but, the areas that are made accessible, such as through Pacific Rim National Park, have experienced many problems in accommodating the large numbers of people, with resulting disastrous effects on both the local ecology and the recreational experiences of many of the visitors.

Before leaving the topic of land-based coastal recreation, it should be noted that the provincial parks on the coast of the Strait of Georgia are among the heaviest used in the entire system. Campgrounds are full throughout the summer at parks such as Miracle Beach, Rathtrevor and Bamberton, even though they have the highest per night fees in the province. Major activities are those associated with camping, building fires, for example, and beach and water-oriented activities.

Conclusion

In this paper an attempt has been made to draw together the many diverse threads that comprise recreational use of the coastal zone on the Pacific coast of Canada. The range is vast. Luxury cruisers, complete floating hotels, leisurely ply the waters of the Inside Passage for those little interested in close encounters of the environmental kind, while others pound the West Coast Tail looking for wildlife, wilderness and physical challenge. All along this spectrum the demand for activities appears to be rapidly increasing, both from residents and non-residents alike. The main technical report of the Canada-British Columbia Travel Industry Development Subsidiary Agreement indicates the highly significant role that marine recreational activities will play in the future as tourist attractions (Canada and British Columbia, 1979).

Despite the disparate nature of the activities and infrastructure required to support them, all are vulnerable in the future to threats from the same kinds of forces. First, all these activities are, to a greater or lesser extent, dependent upon the maintenance of key elements in the coastal environment for which there is keen competition from other resource-based industries. For example, the sport fishery is dependent upon there being a reasonable opportunity for, and probability of, the sport angler catching salmon. Without the salmon there would be no associated recreational activity, nor business built to serve it. A corresponding decline would be felt through the pleasure boating industry. Yet, as has been mentioned earlier, salmon stocks are seriously depleted. Following proposals from the Department of Fisheries and Oceans to severely limit sport fishing to conserve stocks, the sport fishermen suddenly became aware of their minimal influence on the way in which the resource was managed and allocated. Now a strong sport fishing lobby has managed to considerably strengthen the voice of the recreational fishermen in managing the resource.

The point to be made here is that all the coastal recreational activities are dependent to some extent for their existence and enjoyment upon resources over which they have no control. Until the recreational opportunity becomes so badly impaired as to be negligible, this resource dependency is not always obvious and, hence, of little concern to the recreationists or those responsible for management. A recent letter to the Victoria Times-Colonist provides another example:

> On a recent trip up the Inside Passage, I was appalled at the destruction from logging operations... the area across from Alert Bay has undergone a huge clear-cutting. It appears the logging companies are given far too much leeway to clearcut, and no visible attempt is made to prevent erosion. Numerous cruise ships travel the Inside Passage and each year the view deteriorates. We should try to keep this passage as presentable as possible if we want to maintain the tourist industry. (June 1, 1984 p. A4).

A major component for many of the coastal recreational activities is the aesthetic pleasure derived from the scenery, yet, indiscriminate clear-cutting along the coast is causing vast changes in the landscape. How far can aesthetic satisfactions be compromised before the activity is no longer pleasurable becomes the relevant question, especially, as noted in the letter above, with reference to the tourist industry. This point has been discussed in more detail elsewhere (Dearden, 1983a, 1983b).

It is obvious that there are many competing and conflicting demands for resources of the coastal zone. However, despite the widely acknowledged significance of the area as a recreational resource, recreation is seldom one of the factors taken into account in managing coastal resources. As a result, the recreational potential of the coast often is impaired as an externality of the operation of other resource users. This will continue to be the case unless recreationists and the recreation industry join together to stake a claim in the management of the coastal resource. The most recent conflict to rise in this regard is between the rapid growth of aquaculture in many sheltered bays along the coast which totally foreclosed recreational use of these areas.

The second major cause for alarm in the future relates, not to the external threats, but to the threats from within. The phenomenal rise in popularity of outdoor recreation over the last three decades has been felt very strongly in the coastal zone. As essentially a linear feature, there is a greater concentrating effect of recreationists than is generally felt in purely terrestrial situations. In short, there are more people doing more recreational activities on the coast than ever before. In all likelihood, there will be more next year, and even more the year after that. This creates many resource management problems, leading to degradation of the biophysical habitat and conflict between overcrowded recreationists. Solving these twin problems of conflict between recreational and other uses of the coast and conflicts generated by rising numbers of recreationists constitute a major challenge in coastal zone management on the Pacific coast in the future.

References

Almack, R. 1983. 'West Coast Hiking Trail, Pacific Rim National Park', *Outdoor Reports*, August, p. 11.

Canada and British Columbia, 1979. *Technical Reports of the Province of British Columbia Tourism Development Strategy*. Ottawa and Victoria: Canadian Office of Tourism and B.C. Ministry of Tourism.

Cooke, K. A. 1977. *Transient Boating in the Strait of Georgia*. Vancouver, M.A. Thesis. University of British Columbia.

Daniels, A. 1986. 'Berth Space Questioned', *Vancouver Sun*, March 15, p. D1.

Dearden, P. 1983a. 'Tourism and the resource base', in P. Murphy (ed.), *Canadian Tourism: Issues and Answers for the 1980s*. Western Geographical Series, Department of Geography, University of Victoria, pp. 75–90.

Dearden, P. 1983b. 'Forest harvesting and landscape assessment techniques in British Columbia', *Landscape Planning* 10, 239–253.

Dearden, P. 1986. 'Desolation Sound Marine Park, British Columbia', in J. Lien and R. Graham (eds.), *Marine Parks and Conservation: Challenge and Promise* 2, 157–167. Toronto: Canadian Parks and Wilderness Society.

Dearden, P. 1987a. Oceanic wilderness users: A case study of soil boat charters to South Moresby Island, British Columbia. Paper presented to Oceanic Wilderness Seminar. World Wilderness Congress, Denver, Colorado.

Dearden, P. 1987b. Mobilizing public support for environment: The case of South Moresby Island, British Columbia. Paper presented to the 17th Annual Joint Meeting of the Environment Council of Alberta, Edmonton, December 4. Published in *Proceedings*, 62–75.

Dearden, P. 1987c. 'Marine ecosystem protective designations in British Columbia', in *Heritage for Tomorrow: Canadian Assembly on National Parks and Protected Areas*

R. C. Scace and J. G. Nelson (eds.), Ottawa, Environment Canada-Parks, Vol. 3, 123–146.

Department of Fisheries and Oceans, 1987. *Sport Fishing Data*. Vancouver, DFO.

Duffus, D. 1988. Whale-watching on the Pacific Coast of Canada. Ph.D. Dissertation. University of Victoria, Victoria, B.C.

Dufour, P. 1985. 'Cruises to Alaska inject millions into economy', *Victoria Times-Colonist*, August 7, p. B1.

Ernst and Whinney, 1980. *An Evaluation of the Tourism Potential of the Scuba Diving Industry in British Columbia*. Ottawa and Victoria: Canadian Office of Tourism and B.C. Ministry of Tourism.

Harrison, M.C. 1979. *Resident Boating in Georgia Strait*. Vancouver, Environment Canada.

Montgomery, G. 1981. *An Evaluation of the Tourism Potential of the Cruise Ship Industry of British Columbia*. Ottawa and Victoria: Canadian Office of Tourism and B.C. Ministry of Tourism.

Pearse, P. H. 1982. *Turning the Tide: A New Policy for Canada's Pacific Fisheries*. Vancouver, Department of Fisheries and Oceans.

Rhodes, A. 1979. *A General Survey of Non-Resident Boating Populations in the Gulf of Georgia*. Vancouver, Department of Fisheries and Oceans.

Philip Dearden
Department of Geography,
University of Victoria,
Canada

11. The recreational use of the Norwegian coast

The first to use the Norwegian coast as a recreational area were the passengers on board cruise ships. Coming from England and America, these tourists started to visit the fjords of Norway around the turn of the century: a culminating point of many cruises was a visit to Cape North. It was, however, the coastal scenery and the life-style of the people, which attracted most of these tourists, who used to make short detours up the valleys to the mountains along the western coast. To facilitate their stay ashore, hotels were built at the head of many of the fjords; some of these are still operating today.

The coast of Norway, with its infrastructure of roads and hotels has, then, a long tradition regarding this kind of recreational use. This use however, was originally limited to the upper classes. For the people living along the coast, local resources, such as fishing and timber, provided them with enough food and materials to be self-sufficient. Between the World Wars, a few people with high economic standards built recreation places at the seaside. The habit of staying at a bath or a spa never found its place at the Norwegian coast, although there were a few hotels for summer recreation on the Oslofjord and Skagerrak coast (Figure 1). At the same time, the recreational use of some sandy beaches by people from nearby towns started on a small scale. The use of coastal areas for recreation became more common after the Second World War.

Types of recreational use

Their prerequisites and distribution

There are different types of recreational use of the Norwegian coast: 1) fishing, 2) bathing, 3) sunbathing, 4) strolling, 5) birdwatching, 6) diving, 7) sailing, 8) windsurfing, 9) boating, 10) cruising, 11) camping, 12) staying in privately owned huts or cabins, 13) staying in camping huts or shanties (previously fishermen's quarters), 14) staying in apartment huts or cabins, and 15) staying in boarding houses or hotels.

The recreational use of a coastal area is very often a combination of two or more of the activities presented in the above list. Some of these activities require a short comment, but more important than a description, are the comments on the 'prerequisites' involved and also the lack of some of these prerequisites.

Fishing, either from boat or land, is found everywhere along the coast except for a few places where industrial pollution has spoiled the environment. Bathing takes place from rocky shores and boats, as well as from sandy beaches. Birdwatching is localized at places along migrant routes and at breeding sites in rock walls.

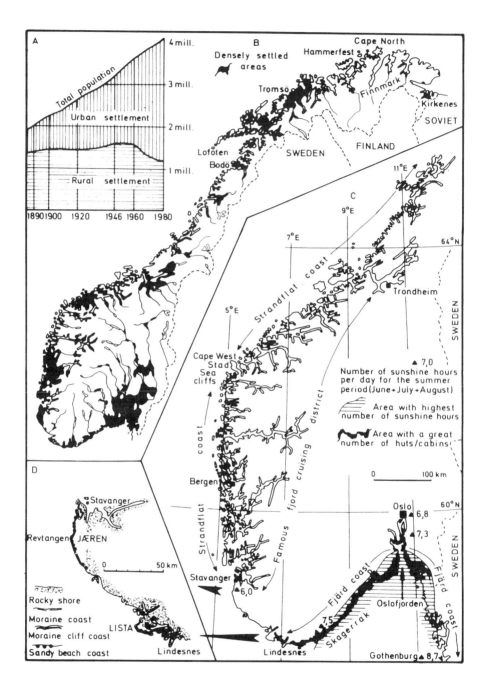

Figure 1. Temporal (A) and spatial (B) distribution of population in Norway. Coastal landscapes in southern Norway (C) and the Jæren – Lista districts (D).

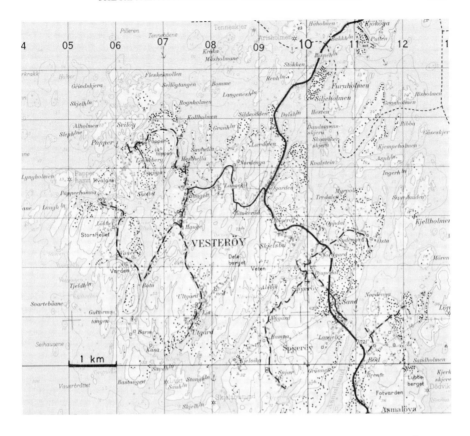

Figure 2. Section of the topographical map showing the fjärd coast and the density of coastal recreational settlement (black squares: huts and cabins) on the southeastern side of the Oslofjord. (Printed by permission of the Norwegian Mapping Authority).

Strolling along the sandy beaches is popular, but as there are only a few long sandy beaches, strolling also takes place along stoney beaches and along the foot of the moraine cliff coast, as well as along parts of the rocky shore of the fjärd coast.

Boating and sailing can be a weekend habit close to the harbouring marina, but so also can holiday 'strolling' along the coast, with stops in sheltered night harbours in the skerry zone.

Some cruise ships still find their way to Norway and the 'Coastal Express' from Bergen to Kirkenes close to the Soviet border, offers cruises. Many camping sites are found along the coast, offering an inexpensive opportunity for staying compared with staying at hotels. Although privately owned huts or cabins have a wide distribution, it is in the area around the Oslofjord where most of them are found (Figures 1C and 2). Staying at apartment huts and cabins is a new form, found only in a few places, as is staying in the shanties of the Lofoten Islands, used by fishermen during the first half of this century.

Geomorphological prerequisites

The geomorphological setting is a main prerequisite for recreational use of the coastal area. The topography along the Norwegian coast may be divided in three types: either as steep sea cliffs, as more or less steep or precipitous fjord-sides, or as the uneven, rugged topography of the fjärd and strandflat coast (Klemsdal, 1982). The last one is most important, as its rugged topography presents a very indented coastline. According to the Norwegian Mapping Authority's measurements on 1:50 000 maps (and some few 1:100 000 maps), the coastline of the mainland of Norway is 21 111 km, and that of the islands 31 958 km, giving a total coastline of 53 069 km. The straight line along the outermost islands from the Swedish to the Soviet border is 2650 km. The total number of islands is 53 789. These figures indicate the intricate coastline of *the skerry zone* with its headlands and bays, inlets and fjords, and a network of islands and sounds.

This very broken and indented coastline with numerous islands, settles the playground for the activities mentioned in the list. The sheltered waters of the skerry zone are suitable for fishing, bathing, sunbathing, sailing, windsurfing, boating, birdwatching, and diving, either for weekend enjoyment or for longer holiday occupation. The skerry zone is, therefore, most important regarding recreational use.

The geomorphology of the sea cliffs, the fjords, the strand-flat, and even the fjärd coast, is a vital element of the coastal scenery and landscapes which have attracted holiday-makers.

Finally, the geomorphological prerequisites include the local setting of landforms and the fact that the shore zone of Norway is mainly a rocky shore intermingled with stoney beaches (Klemsdal, 1979). Sandy beaches are few and the number of long beaches is restricted and are mainly found at Lista and Jæren (Figure 1). Beside these, a number of short pocket beaches are found in between the rocky shores of the fjärd coast. Clayey beaches (Klemsdal, 1979), resembling the salt marshes, are locally distributed at the outlet of some rivers and at the fjärd coast, where the clayey plains of the joint valleys gradually slope into the sea.

This distribution of bedrock and surficial deposits in the shore zone is usual for a former glaciated area. The small number of sandy beaches and extensive rocky shores has made it natural and necessary for the people to turn to the ice-polished and smoothed landforms of the skerry zone of the strandflat and the fjärd coast. The people's use of the coastal area is adapted to this prerequisite.

Climatological prerequisites.

Recreational use of the coast is normally related to fine, warm, and sunny weather. What is meant by fine weather is highly individual and depends on the recreational activity involved. The long north-south extension of Norway results in sensible variations in climate. Precipitation and temperature data in general reveal fairly good conditions for summer recreation along the coast. In addition to this, it is necessary to deal with a local climate effect. Along the southeastern coast, the

Skagerrak, and southern Oslofjord coast, the number of sunshine hours is at its greatest (Figure 1C). The convective precipitation during warm summer days occurs somewhat inland from the outer coast. This means that the outer coastal districts of southern Oslofjord and along the Skagerrak coast, have more sunshine and a more favourable climate for recreation than the inland areas. The meteorological situation connected with convective precipitation also produces a wind regime called the 'summer sea breeze'; a prerequisite for windsurfers and sailors.

The northerly location of the Norwegian coast does have a drawback: the sea temperature. During normal warm summers, the sea temperature in the southern Oslofjord district and along the Skagerrak coast is approximately 18 °C with lower values northwards along the coast. Bathing is then localized to the southeastern and southern parts of the country and to places along the coast where the water temperature rises as a result of special, sheltered, or shallow conditions. Locally, these may even be found in the northern parts of the country.

Biological prerequisites

Rich resources of fish in the coastal waters have attracted people for leisure fishing for years. Along the coast, there are many islands and rock walls of sea cliffs where birds have their breeding places. Migrant birds moving along the coast, have their own resting places for feeding, as along the sandy beaches of Lista and Jæren, where seaweed and algae are washed ashore, or in the productive shallow waters of the clayey beaches of the fjärd coast.

The vegetation in the coastal zone is closely related to soil and climate conditions. The coast of Oslofjord and the northern part of the Skagerrak coast has a mixture of spruce, pine, and deciduous trees (Moen, 1987). Except for the narrow treeless zone further out, pine dominates the outer parts of the coast. Along the southern part of the Skagerrak coast, temperate deciduous forest with a marked element of oak, is found. Sand-dune vegetation fringes the sandy beaches. On the strandflat from Stavanger to Bodø, open heathland (*Calluna* moor) vegetation prevails. North of Bodø, Boreal vegetation occupies the coastal zone to Hammerfest, with coniferous trees in the south and birch in the north. North and east from Hammerfest, low alpine vegetation reaches the sea level.

Social prerequisites

The rebuilding of Norwegian society after the Second World War has brought many changes. Less working hours per week (Svalastog, 1988): 48 h in 1945, 45 h from 1959 and reduced to 42.5 h in 1968 and 40 h in 1976; and at the same time longer statutory vacations per year (3 weeks from 1947 to 4 weeks from 1964), increased available leisure time and the possibilities for recreation. Increases in wages were also stimulating factors for recreation. A great number of huts and cabins were built during the two first decades after the war and the habit of spending vacations along the coast became popular.

As the population in the post-war period grew (Figure 1A) simultaneously with the migration of people from the rural districts and the northern parts of Norway, a concentration in the larger towns (Oslo, Bergen, Trondheim, and Stavanger) and the smaller towns in southeastern Norway took place (Figure 1B). Today, approximately 75% of the population lives within 15 km of the coastline, with strong concentrations around the Oslofjord and the main towns. The trend of owning a hut or cabin placed a heavy pressure on the coastal areas. This pressure reached a level where the authorities had to introduce a temporary law in 1965 against the building of new huts and cabins within 100 m of the shoreline. This prohibition was included in the Shore and Mountain Planning Act of 10 December 1971.

The increased economic standards made the people more mobile and the law against new huts and cabins has resulted in a strong interest in coastal waters. The number of cabin cruisers, more than 24 feet long, increased in the 1970s, and the need for marinas became strong. As a result, boating and 'strolling' along the skerry zone increased the pressure on the coast, especially along the Oslofjord and Skagerrak coast where ports and spontaneous night harbours have been heavily used.

Management

A growing use of the coastal zone for recreation, with a strong pressure on the coastal environment in the vicinity of the main towns and the Skagerrak and the Oslofjord coast, has made it obvious that a policy of management was needed. The result of the increased numbers of huts and cabins in the post-war period, forced the authorities to take care of the coast. The Council of Open Air Recreation for the Oslofjord area bought areas of land and made arrangements with other owners to have certain islands and parts of the coastline free from huts and open for everyone to use. The drawback of the increased use was a deterioration of the environment, especially in the vegetation around popular sheltered night harbours. An intense use of the coastal zone through boating, has forced the authorities to restrict the use of certain places in the skerry zone on the southwestern side of the Oslofjord.

Only one national park in Norway has a small part of its border along the coast. This means that no national park has been established in relation to coastal environment conditions. There is, however, a great number of nature reserves along the coast and many of these have been established to protect the environment for birds. A lot of breeding places have been turned into bird sanctuaries.

The Council of Open Air Recreation of Jæren has long worked for the protection of the local coastal zone. The coastal types in that area (Figure 1D) are a moraine coast and an inactive moraine cliff coast with stoney beaches interrupted by stretches of sandy beaches and dunes backed by agricultural land. This open coastal landscape close to the densely populated area of Stavanger, is a most valuable recreation area. In 1971, the Protected Landscape Area of the Jæren coast was established, but due to competition with agriculture, the area is rather narrow. The protected landscape area is heavily used for strolling, sunbathing, and birdwatching, but it is not used much for bathing, as the water temperature is low. In 1987, the Protected Landscape Area of the Lista coast was established in a similar environmental setting.

Concluding remarks

The recreational use of the Norwegian coast closely depends on the combination of physical and biological, as well as on cultural and social prerequisites.
   The coastal landscapes along the Norwegian coast (Figures 1B, 1C and 1D) and their recreational use can, thus, be summarized as:

Sea cliff landscape. Sea cliff landscape is found on the northern parts of the west coast; around Stad and in Finnmark, in the northeast. The sea cliff scenery of Cape North not only attracts tourists on board cruise ships, but also tourists arriving by car from all over Europe. Cape West on Stad does not enjoy the same international reputation.

Strandflat landscape. As a zone, the strandflat stretches along the coast from Stavanger in the southwest to western Finnmark in the northeast. The coastal environment with its treeless heather vegetation and open views with numerous rocky islands, presents the scenery of the skerry zone as varied and attractive to visitors who find travelling in sheltered waters easy and pleasant. The main recreational use is touring, but many other activities, like sailing, windsurfing, boating, and strolling take place. Bathing, however, is hampered by low water temperatures. The recreational use of the strandflat landscape is closely related to the pressure of people and the areas around the main towns are heavily exploited.

Fjord landscape. The fjords follow inland from the strandflat and are therefore also found from the southwest to the northeast of Norway. The fjord scenery, the steep fjord sides and small settlements have drawn many tourists to this kind of landscape. Tourists come by cruise ships or by cars and buses. Activities connected with the use of the coast, e.g. boating and sailing, windsurfing and bathing, are, however, of minor importance in the fjord districts.

Sandy beach landscape. The sandy beaches of Lista and Jæren with their bordering dune areas, have a great appeal for people. Although the main drawback of the Norwegian sandy beaches is the water temperature, beaches, however, are used for sunbathing, strolling, and birdwatching.

Moraine coast landscape. The moraine coast, with or without the fossil moraine cliff along the coast of Lista and Jæren, has stoney beaches bordered by agricultural land. This environment does not attract many people to come to the shore for recreation, and the stoney beaches are only used for strolling.

Fjärd landscape. The fjärd landscape is found from the Swedish border in the southeast to the southernmost point of the country at Lindenes. Here the environment is a combination of the skerry zone with ice-polished rocky shores, sheltered waters, an open pine forest mixed with deciduous lik oak, and a warm climate during summer with plenty of sunshine. All these prerequisites result in favourable conditions for recreation and vacation, with excellent possibilities for many different activities. The coast of Oslofjorden and the Skagerrak, being close to the densely populated areas of southeastern Norway, turns out to be the main area for recreational use along the Norwegian coast.

References

Klemsdal, T. 1979. Kyst-, strand- og vindgeomorfologi. Forslag til terminologi. (Coastal-, shore/beach- and eolian geomorphology. Proposal for terminology.) *Norsk geogr. Tidsskr.* **33**, 159–171.
Klemsdal, T. 1982. Coastal classification and the coast of Norway. *Norsk geogr. Tidsskr.* **36**, 129–152.
Moen, A. 1987. The regional vegetation of Norway; that of Central Norway in particular. *Norsk geogr. Tidsskr.* **41**, 179–225.
Svalastog, S. 1988. Tourism in Norway's rural mountain districts twenty-five years after the mountain planning team's report. *Norsk geogr. Tidsskr.* **42**, 103–120.

Tormod Klemsdal
Department of Geography,
University of Oslo,
Norway

12. Patterns and impacts of coastal recreation along the Gulf coast of Mexico

Introduction

Much of Mexico's shoreline has been transformed into a recreational cultural landscape. Segments of the Pacific coast and (more recently) the Caribbean coast have undergone such extensive cultural and physical modification by tourism that a recent regional study of Mexico broke them out as a separate 'nation' of 'Club Mex' (Casagrande, 1987). Included in Club Mex are the Pacific enclaves of Mazatlán, Puerto Vallarta, Manzanillo, Ixtapa, Acapulco, and Puerto Escondido and the Caribbean enclave of Cancún/Cozumel. Coastal tourism accounts for approximately 45% of total tourism in Mexico, which translates to about $700 million in (1983) revenues (Merino, 1987), and the Club Mex enclaves are the primary destinations of most tourists, both international and national. Recognizing the touristic value of its shores, Mexico devoted 93 percent of its (1982) investment in tourism – $370 million – to coastal infrastructural development. Major resort complexes were developed in the 1970s by the Mexican government at Ixtapa-Zihuantanejo on the Pacific coast (Reynoso y Valle and de Regt, 1979) and at Cancún on the Caribbean coast (Collins, 1979).

Conspicuously absent from any discussions of international coastal tourism in Mexico is the Gulf Coast, apparently because of less-than-suitable physical conditions. When the Mexican government decided (in the late 1960s) to develop the infrastructure for an Atlantic resort to counterbalance the numerous Pacific resorts, its computer selected the Cancún site in Quintana Roo (Dunphy, 1972). The beaches along the Gulf of Mexico coast were considered to be 'physically unsuitable for major tourism development' because of a combination of climatic, water quality, and beach quality factors (Collins, 1979). Climatically, negative aspects for tourism include the two dozen or so annual winter cold fronts (nortes) which penetrate to the southern Gulf (and thus made winter tourism opportunities unreliable, especially in comparison to Acapulco and Cancún) and the humid, rainy, and overcast summers (especially in the states of Veracruz, Tabasco, and Campeche). Hurricanes which enter the Gulf tend to track either westward or refract to the north and thus pose only a threat mostly to the sparsely developed Tamaulipas coast. Water quality along the Gulf is lower than the Pacific and Caribbean coasts because of high sediment inflow (especially along the southern and western rim of the Gulf), high nearshore turbidity because of a virtual absence of protective offshore reefs, and localized pollution (especially near Veracruz and Tampico) which tends to attract sharks. Beach quality is also somewhat inferior because of a high amount of fine sediments along the Veracruz coast and a high shell content along the north Yucatán coast.

Recreational development along Mexico's Gulf Coast

In spite of the perceived adverse physical conditions, domestic demand for beach recreation opportunities has nonetheless transformed many segments of the Gulf Coast into distinctively recreational seaside landscapes. A tradition of beach recreation in Mexico is documented at least to the 1780s, when wealthy Spanish ranchers of the lower Río Grande Vally migrated to Matamoros beaches in the hot summers (WPA, 1940), but it did not become popularized until the present century (Passariello, 1983). Significant coastal recreational development did not take place until the provision of beach access, mostly in the form of highways built during the 1930s and 1940s. As access was facilitated and seaside recreation became more popular among the Mexican middle classes, coastal development became closely correlated to the recreational hinterlands, or market areas, of the various suitable reaches of shoreline (Meyer-Arendt, 1987b). The simplest form of beach resort may cater only to a nearby population center, while shorelines highly physically attractive in terms of water clarity, beach sands, and prevailing climate may attract recreationists and tourists from a hierarchy of stacked (or nested) local, regional, national, and international hinterlands. Although Mexico's high-amenity Pacific and Caribbean beaches attract beach lovers from the entire spectrum of recreational hinterlands, recreational development along the physically less attractive Gulf Coast beaches has been primarily a result of domestic tourism, at levels ranging from local to regional to national (Meyer-Arendt, 1987c).

Based on field surveys conducted in 1985, a total of 67 recreationally utilized sites along the Gulf Coast (between the Río Grande and Cabo Catoche) are identified (Figure 1). These sites can be classified either as 'beach resorts' with available accommodations or as 'recreational beaches' in which at least minimal services existed. The resort category can be subdivided on the basis of size and quality of lodging facilities, and 15 sites (22% of the total) can be classed as major resorts (at least one hotel with over 25 rooms) while 21 sites (31%) are considered minor resorts (with available accommodations). Recreational beaches (31, or 46% of total) include settlements or beaches utilized by recreationists in which at least minimal services (e.g. soft drinks) are available. The distribution of the sites is characterized by three major clusters, largely a function of relative proximity to the urban areas which constitute their recreational hinterlands. Almost half of all sites (46%) are contained within two clusters in the state of Veracruz, and recreationists are drawn both from the city of Veracruz and from Mexico City. A third major resort cluster along the Gulf Coast (containing 25% of the total sites) is in Yucatán, proximate to Mérida. Smaller, less well defined clusters are associated with less populous or more distant local hinterlands such as Villahermosa, Tabasco. Even South Padre Island, Texas draws about 3% of its visitors from Mexico, mainly from a Monterrey hinterland (Myers and Hodges, 1983).

Although the hinterland concept best explains the general distribution of the coastal sites, the specific site location is in large part a function of the local physical environment. A sand beach is the prime recreational attraction at each of the sites, although structural development may take place on a nearby rocky headland or across a lagoon from the beach (as is the case with 10% of the sites).

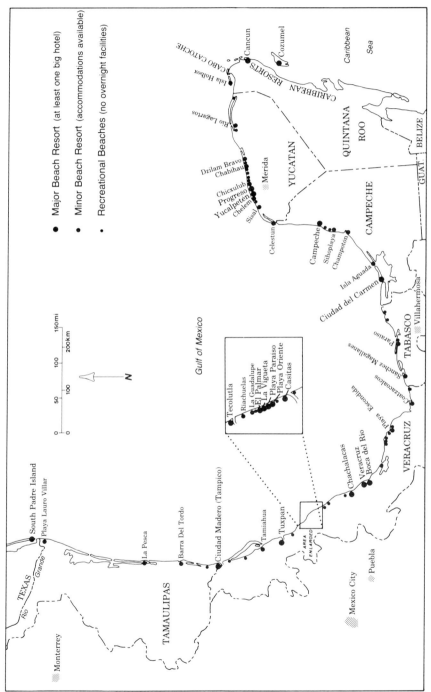

Figure 1. Beach resorts and recreational beaches along the Gulf Coast of Mexico (Meyer-Arendt 1987a) (from field surveys by author).

Isolated pocket beaches comprise a small clustering of sites along the Tuxtla Mountains coast near Playa Escondida. River mouth locations comprise a third (22) of all sites, and wide beaches (often widened artificially because of sand-trapping jetties) are generally found at these sites. Barred freshwater river mouth lagoons (la barra) are especially popular for bathing by families, and 14 (21%) of all sites fell into this category. In the absence of freshwater inflow (such as in drier northern Mexico and along the carbonate Yucatán peninsula), however, swimming in brackish or saline lagoons is not as popular. Non-sandy shorelines are also generally avoided by recreationists, unless perhaps a scenic view is provided. This pattern is aptly demonstrated by the minimum of coastal development along the scrubby volcanic coast of north of Chachalacas, Veracruz and along the seagrass debris shoreline near Dzilam de Bravo, Yucatán.

Because of a combination of factors including locations of the population centers that comprised the recreational hinterlands, distance and accessibility to the coast, and local physical attributes, recreational development along Mexico's Gulf Coast is concentrated in the states of Veracruz and Yucatán.

Coastal recreation in Veracruz

On a national scale, the first major development of Mexico's coastline is traced to the post-Revolution boom years of the 1920s. Automobiles and buses were becoming more common, especially among the upper and upper middle classes of Mexico City. As a result of the increased popularity of seaside recreation, in 1927 the old wagon road from Mexico City to Acapulco was graded into a highway, thus stimulating the first touristic boom at that sleepy Pacific colonial port (Cerruti, 1964). As the 400-km trip required about 24 hours of travel, however, entrepreneurs (with the proper political connections) shifted their attentions to the Gulf of Mexico. Two main areas within the state of Veracruz were soon earmarked for recreational development: the environs of the city of Veracruz and a coastal stretch extending from Tecolutla to Casitas.

The port of Veracruz had been the Atlantic gateway to Mexico since Cortez landed there in 1519, and in 1872 Mexico's first railway linked the port with Mexico City (West and Augelli, 1976). Its steamy malarial setting prevented the city from being any more than an entrepôt (Arreola, 1980) until the popularization of sea-bathing in the 1920s. Traditionally, the Veracruz elite seasonally fled to the cool 'hill stations' of the Sierra Madre for their recreation. Following the 1927 boom in Acapulco, the first major resort hotel along the Gulf Coast – the Hotel Mocambo – was built a few miles south of town at what has become the recreation center of Mocambo Beach (Figure 2). Veracruz proper is flanked by two beach strips: 1) a North Beach (near the port facility as well as the shark-infested point of sewage discharge into the Gulf), and 2) a hotel-lined South Beach. The direction of urban growth is toward the south, and the shorefront between South Beach and Mocambo Beach is rapidly filling in with modern hotels, condominiums, and exclusive subdivisions. The rivermouth town of Boca del Río, south of Mocambo, is presently undergoing conversion from a fishing village into a fashionable upper

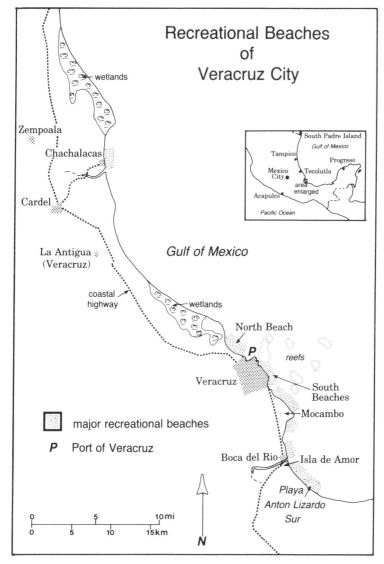

Figure 2. Recreational beaches of greater Veracruz.

middle class seaside resort, and several condominiums have been constructed. The working classes of Veracruz tend to frequent the beaches of Chachalacas, a popular rivermouth resort 16 km north of the city, which is also presently witnessing much hotel and restaurant construction.

Extensive shoreline modification has taken place in the Veracruz area, yet little of it can be directly attributed to coastal recreation. Offshore natural reefs dampen the effects of incoming wave energy and shoreline erosion has been minimal in most of the region. Shoreline modification began as a result of harbor improve-

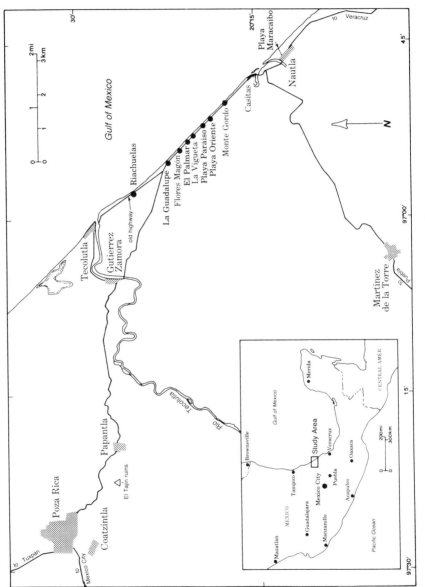

Figure 3. The Tecolutla-Nautla coastal strip.

COASTAL RECREATION ALONG THE GULF COAST OF MEXICO 139

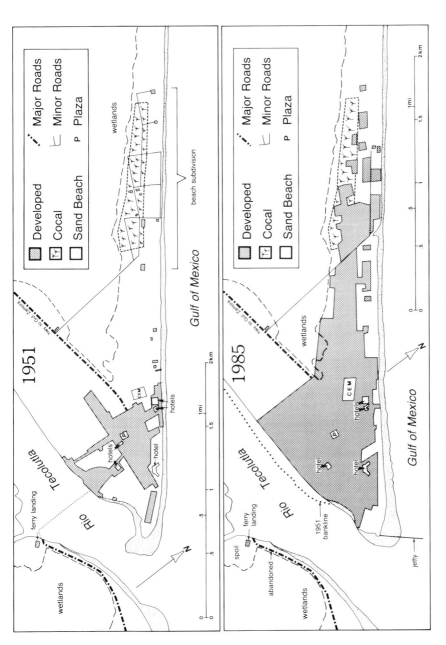

Figure 4. Landuse changes at Tecolutla, 1951–1985.

ments in the late 1800s, when the first breakwaters and groins were built (Gutierrez-Estrada et al., 1988). As the city gradually expanded southward from its walled core, low areas near the shoreline became filled and 'reclaimed', and a low seawall extends for several km. Along one short erosionary stretch south of Mocambo Beach, a groin field was constructed to protect the coastal highway. The construction of rivermouth jetties at Boca del Río (for navigation purposes) has resulted in updrift beach widening and provision of an attractive recreational resource, which may partly explain the recent recreational development boom at that location.

The second cluster of coastal development in Veracruz occurs along the coastal strip between Nautla and Tecolutla, approximately 200 km north of the city of Veracruz (Figure 3). In this region, the recreational hinterland is primarily Mexico City slightly less than 400 km away, and recreational development followed the provision of highway linkages in the early 1940s (Ramirez, 1981). Tecolutla, a small fishing port at the mouth of the Río Tecolutla, soon became recreationally discovered. Promoted as the closest beach to Mexico City in terms of time and distance, Tecolutla was envisioned by speculators as an 'Acapulco East' and three large hotels were built. But significant demand for Gulf Coast resorts never developed, and following improvement of the Acapulco highway in 1955, most new coastal development in Mexico became concentrated along the Pacific coast.

Tecolutla grew slowly between the 1940s and the 1970s. Highway relocation in 1970 left the town isolated from through traffic formerly dependent upon the ferry across the Río Tecolutla. Nonetheless, by the latter 1970s, seasonal recreational demand led to renewed vigor in the resort. The three original hotels, still the major ones, are now filling to capacity during peak periods (notably Holy Week and Christmas), and the construction of new (but modest) lodging facilities has been stimulated. Summer homes, too, are being built in increasing numbers along the beachfront, and a formerly distant beach subdivision (fraccionamiento) has now fused with the main settlement (Figure 4). Most new construction in the area is not in the old resort of Tecolutla proper, but further south along the coast between El Palmar and Casitas, where several medium high-rise resort hotels have been built in recent years (see Figure 3).

The small resorts of Tecolutla and Casitas are situated at rivermouths and thus benefit from high fluvial sedimentation and beach accretion (Self, 1977). The mouth of the Río Tecolutla was jettied in the late 1960s to minimize maintenance dredging for navigational purposes, and the jetties have accelerated the accretion of the updrift beaches. As at Boca del Río south of Veracruz, the widening beaches have enhanced the recreational appeal of Tecolutla. Erosion is a greater problem with distance from rivermouths and some of the newer recreational developments such as Playa Paraíso and Playa Oriente are beginning to be threatened by shoreline retreat. So far, no erosion control structures have been built, however.

Coastal recreation on Yucatan's north shore

Recreational usage of Yucatán beaches dates to the establishment of Progreso as

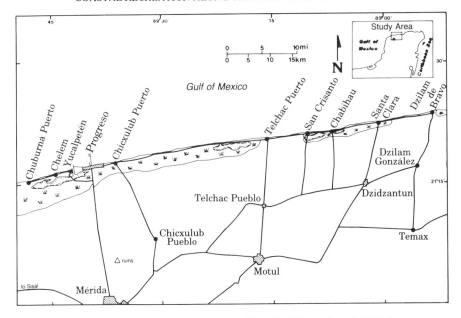

Figure 5. The barrier coast of northern Yucatán (Meyer-Arendt 1987a).

Yucatán's port and its subsequent rail linkage with Mérida in 1881 (Figure 5) (Meyer-Arendt, 1987a). Although the prime function of the railroad was to facilitate the export of sisal (henequén), passenger service was provided to transport 19th century Meridanos to the Progreso beaches. The wealthier families began to build summer residences (casas veraniegas), and by 1907, Progreso boasted of three hotels and a large beachfront recreational complex (Frías and Frías, 1984). By 1928, the automobile age had arrived in Yucatán, and the Mérida-Progreso highway was paved. Increasing numbers of summer homes were built along Progreso's shorefront, behind the promenade (malecón) and along the beach toward the east. Although the only paved road during the 1930s and 1940s was the Mérida-Progreso highway, graded roads extended toward Chicxulub Puerto and Chelem and sand roads ran onward to Dzilam de Bravo and Chuburná Puerto. Roadbeds were built across the coastal lagoon at various locations to provide better port access to inland settlements. This facilitation of access, coupled with growing usage of motorized vehicles, expanded the potential for coastal recreation.

The pattern of recreation use was still primarily one of day use in 1945, although the trend of vacation home ownership was beginning to expand beyond Progreso, initially to Chicxulub (Martínez, 1945). As early as 1948, Progreso and Chicxulub were being welded together by a ribbon of beachfront housing (Figure 6). To the east, summer homes were colonizing the coconut plantations (cocals) and barren beach ridges. If Progreso is seen as the core area of beach recreation, this zone represents a recreational 'frontier' marked by summer homes built in advance of the provision of basic services. Smaller settlements along the north Yucatán coast such as Telchac Puerto, San Crisanto, and Chabihau also became recreational nodes

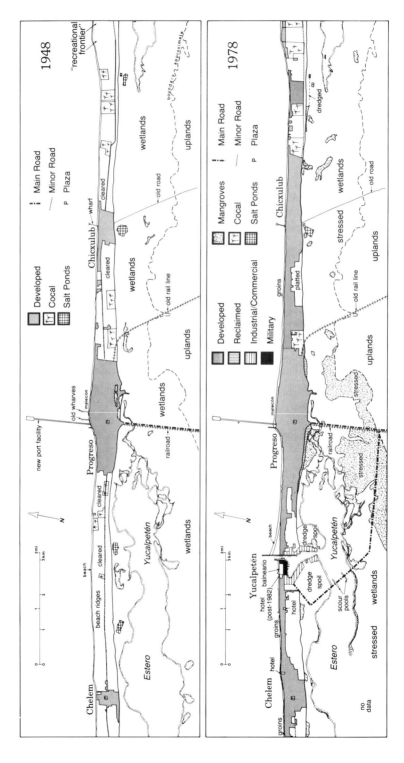

Figure 6. Landuse changes at greater Progreso, 1948–1978 (Meyer-Arendt 1987a).

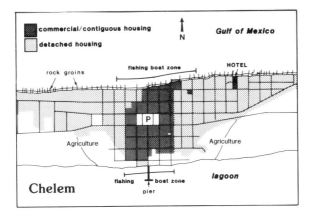

Figure 7. Urban morphology of Chelem (Meyer-Arendt 1987a) (schematic from oblique photos taken by author, 1984).

from which summer home development spread laterally. Mérida provided the majority of recreationists to this coast, although distal beaches such as Telchac Puerto also drew from secondary urban centers such as Motul and Temax (see Figure 5).

Because of availability of utilities and easy access from Mérida, the Progreso vicinity remained most popular for summer home construction. After Progreso's beachfront filled in with summer homes during the 1950s, Chicxulub became the primary locus of recreational growth. A secondary direction of expansion during the 1950s was westward toward Chelem (Figure 7), which, like Telchac Puerto became a recreational node from which summer home construction spread both to the east and west. Chuburná Puerto presently marks the western limit of the 20-km long contiguous north coast recreational landscape. Toward the east, the shoreline is still relatively pristine, although the 'recreational frontiers' continue to shift eastward from Chicxulub Puerto and outward from the smaller nodal settlements. Within the next several decades, the entire coast from Chuburná Puerto to Dzilam de Bravo will undoubtedly witness extensive beachfront urbanization.

The sandy barrier coastline of northern Yucatán has historically been characterized by slight shoreline erosion, both as a result of normal wave action and also by human interference with the prevailing east-to-west longshore sediment drift. The Progreso waterfront, for example, has historically experienced much local erosion as a result of port construction, and a rare hurricane in 1947 destroyed many beachfront structures. The first human responses to erosion, however, were efforts to improve the beaches for recreationists. In 1964, several rock-and-timber groins (espolones or escolleras), designed by government engineers, were installed fronting the Progreso malecón. In the short term, the espolones proved to be relatively successful in offsetting a trend of beach erosion.

Armoring of the recreational shorefront increased after the construction of the safe harbor (puerto de abrigo) at Yucalpetén in 1968 (see Figure 6). Although the storm-protected safe harbor provided a suitable base for the Progreso fishing fleet,

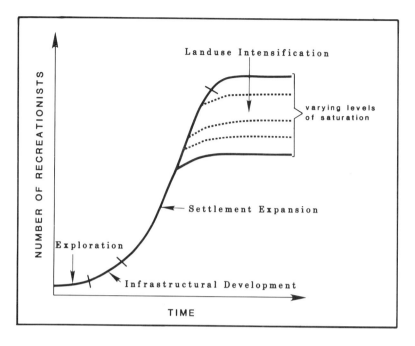

Figure 8. A theoretical model of resort evolution (loosely based upon a model by Butler 1980).

a naval base, and a growing seafood processing industry, the dredging of a channel through the barrier island led to accelerated shoreline erosion downdrift of the jettied entrance. In response to the high rates of erosion (the 1978 map indicates retreat of over 30 m in a 10-year period immediately west of the jetties), widespread unauthorized espolón construction began (see Figure 7). Beachfront lot owners built espolones – eventually extending westward to beyon Chelem – without obtaining either advice on engineering or the legally required construction permits. Approximately 75% of the vacation home properties west of the Yucalpetén jetties presently encroach to within the 25-meter wide federal beach easement as a result of beach erosion, and many seasonal landowners perceive groins as a means of saving their property. As groin construction led to increased downdrift erosion, more groins were built in response, and between 1968 and 1985 the leading edge of espolón construction continued to migrate westward. During a 1984 aerial survey, 178 espolones were noted along the 8.8 km stretch from Yucalpetén to Chuburná Puerto, an average of one every 50 m. Groin density is highest between the jetties and Chelem. As a consequence of this shoreline modification, the attractiveness (and widths) of the beaches diminished. In the early 1980s, officials from Chuburná Puerto formally complained that espolones had increased shoreline erosion within their jurisdiction, and by April 1985 the ban on unauthorized groin construction became actively enforced. Several groins have been removed from the Chuburná ejido beachfront since 1985.

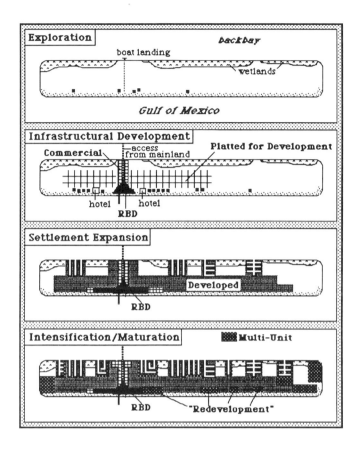

Figure 9. A model of morphologic evolution of a seaside resort.

A Gulf Coast model of recreational development

In recent studies of recreational development patterns along the Gulf Coast of both the United States and Mexico (Meyer-Arendt, 1987b; Meyer-Arendt, 1988), temporal (theoretical), spatial (morphological), and environmental models of resort development were proposed. Theoretically, resort evolution progresses along a classic S-curve model (Figure 8), in which the x-axis represents time and the y-axis measures development levels either by numbers of tourists or numbers of housing units. Progression of resort development along the curve can be described by at least four evolutionary stages – *exploration, infrastructural development, settlement expansion,* and (a levelling off stage of) *maturation* – with the potential of a fifth (*landuse intensification*) if levels of recreational demand remain high enough to warrant higher density development. Each of the evolutionary stages may also be described by a characteristic urban morphology. These patterns of evolving resort morphology may, in turn, be schematically modelled (Figure 9). Modelling of the environmental aspects of resort evolution proved to be less accurate because of the high amount of variability in physical parameters such as winds, wave energy, shore erosion vs accretion, and storm frequency, among others. In the case of erosionary shorelines, the greatest magnitude of structural modification was found to fall more within the settlement expansion stage than in the higher, levelling off, stages.

Whereas the exploration stage is characterized by day use visitation and perhaps a sprinkling of beachfront cottages (see Figure 9), in the stage of infrastructural development a conscientious decision has been made by one or more entrepreneurs to recreationally develop the site. At the beach terminus of the access corridor, a recreational business district (RBD) develops, and one or more hotels may be built nearby. If recreational demand exists, the site will enter a stage of settlement expansion whereby the RBD will expand laterally along the beach, the access corridor will become more commercially developed, and the distal beachfront zones and more interior sites will fill in with recreational housing (mainly with single-family structures while real estate values are still relatively low). Residential canal subdivisions may characterize the backbarrier if conditions warrant and if demand exists. Eventually, all available land will fill in, and the resort may enter a stage of landuse intensification if recreational demand remains high. Developers may replace low density forms of landuse with highrise hotels and condominiums. The level at which a resort reaches a growth plateau represents the stage of maturation.

Summary

Comparison of U.S. resorts with Mexican resorts reveals mostly similarities in patterns of development, although the Mexican Gulf Coast study sites (Tecolutla and Progreso) are still in earlier stages of development. While the U.S. resorts were classified as being in advanced stages of development (either active landuse intensification or having reached plateaus at varying levels of maturation), the Mexican Gulf Coast resorts fell into varying degrees of the settlement expansion stage of the model. A smaller Mexican leisure class, a depressed national economy, perceived adverse physical conditions, and alternative tourism opportunities (e.g. the Pacific resorts for Mexico City recreationists) may all be held accountable for the slow rates of growth along the Gulf Coast. However, development pressures are locally high, as in the vicinities of Veracruz and Progreso. During the stage of settlement expansion along the eroding shoreline west of Progreso, (unauthorized) human modification of the shoreline by groin construction was widespread, and considerable beach degradation has been the result. At the other major recreational shorelines in the state of Veracruz, the rivermouth resorts (e.g. Ciudad Madero, Tuxpan, Tecolutla, Casitas, Chachalacas, and Boca del Río) have benefited from accretion due to both natural processes and jettying of channels.

In view of the Mexican government emphasis on developing international tourism in the Club Mex enclaves along the Pacific and Caribbean coasts, it is likely that the past recreational development trends along the Gulf Coast will continue as outlined above. In Tabasco, the rerouting of the main highway from an inland location to along the barriers (projected for completion in late 1988) may stimulate minor resort development, but growth within the three resort clusters in Veracruz and Yucatán is expected to remain the dominant trend.

Acknowledgements

This material is partially based upon work supported by the National Science Foundation under Grant No. SES-8507500. Any opinions, findings, and conclusions or recommendations expressed are those of the author and do not necessarily reflect the views of the National Science Foundation. Previous fieldwork in 1984 was supported by the Robert C. West Graduate Student Fellowship Fund of the Louisiana State University Department of Geography and Anthropology. Many thanks are extended to Michele D. Meyer-Arendt for the cartographic illustrations accompanying this paper!

References

Arreola, D. D. 1980, Landscapes of Nineteenth-Century Veracruz, *Landscape* 24 (3), 27–31.
Butler, R. W. 1980, The Concept of a Tourist Area Cycle of Evolution: Implications for Management of Resources, *Canadian Geographer* 24, 5–12.
Casagrande, L. B. 1987, The Five Nations of Mexico, *Focus* 37 (1), 2–9.
Cerruti, J. 1964, The Two Acapulcos, *National Geographic* 126, 848–878.
Collins, Charles O. 1979, Site and Situation Strategy in Tourism Planning: A Mexican Case Study, *Annals of Journalism Research* 6, 251–366.
Dunphy, Robert J. 1972, Why the Computer Chose Cancún, *New York Times* 10 (March 5), 1, 27–28.
Frías, B. R. and Frías, B. R. 1984, *Progreso y Su Evolución, 1910–1917*. Editorial El Faro, Progreso, Yucatán.
Gutierrez-Estrada, M., Castro-Del-Río, A, and Galaviz-Solis, A. 1988, Mexico, in *Artificial Structures and Shorelines* (H. J. Walker, ed.), D. Reidel Publishing Co., Dordrecht, Netherlands, 669–678.
Martínez, H. and Victor, M. 1945, Geografía política, demográfica y económica de Yucatán, in *Enciclopedia Yucatanense* (C. A. Echanove Trujillo, ed.). Government of Yucatán. (reprinted 1977), Vol. 6, 455–516.
Merino, M. 1987, The Coastal Zone of Mexico, *Coastal Management* 15, 27–42.
Meyer-Arendt, K. J. 1988, Morphologic Patterns of Resort Evolution along the Gulf of Mexico, paper presented at the 84th annual meetings of the Association of American Geographers, Phoenix, April 7–10, 1988.
Meyer-Arendt, K. J. 1987a, Recreational Landscape Evolution along the North Yucatán Coast, *Yearbook, Conference of Latin Americanist Geographers* Vol. 13, 45–50.
Meyer-Arendt, K. J. 1987b, *Resort Evolution along the Gulf of Mexico Littoral: Historical, Morphological, and Environmental Aspects*. Ph.D. Dissertation in Geography, Louisiana State University, Baton Rouge (reprinted by University Microfilms International, Ann Arbor, Michigan).
Meyer-Arendt, K. J. 1987c, Seaside Recreation Patterns along the Mexican Gulf Coast, paper presented at the 83rd annual meetings of the Association of American Geographers, Portland, Oregon, April 22–25.

Myers, S. J. and Hodges, L. 1983, *The 1983 South Padre Island Tourism Study*, Dept. of Recreation and Parks, Texas A and M University, College Station.
Passariello, P. 1983, Never on Sunday? Mexican Tourists at the Beach, *Annals of Tourism Research* **10**, 109–122.
Ramirez, L. D. 1981, *Tecolutla: Monografía Histórica*. Universidad Veracruzana, Jalapa, Veracruz.
Reynoso y Valle, A. and de Regt, J. P. 1979, Growing Pains: Planned Tourism Development in Ixtapa-Zihuantanejo, in *Tourism – Passport to Development?* (E. de Kadt, ed.), publ. for World Bank and UNESCO by Oxford University Press, New York, 111–134.
West, R. C. and Augelli, J. P. 1976, *Middle America: Its Land and Peoples*. Prentice-Hall, Inc., Englewood Cliffs, N.J.
Works Progress Administration (WPA) 1940, *The WPA Guide to Texas*. Reprinted in 1986 by Texas Monthly Press, Austin.

Klaus J. Meyer-Arendt
Department of Geology and Geography,
Mississippi State University,
U.S.A.

13. Wetlands recreation: Louisiana style

Unlike other US coastal states, Louisiana's coastal zone is not rimmed by attractive beaches, cliffs or bluffs. More than 3.2 million ha of marsh and intermingled water surfaces dominate the landscape. Forty-one percent of The United States' coastal marshes and 25% of the country's wetlands are in Louisiana. It is one of the world's largest and most biologically productive wetlands. This biological resource is a bonanza of aquatic and avian wildlife (Figure 1). As an outdoor resource, the aquatic lowlands serve as a focal point for family-oriented activity.

Louisiana's recreational wetlands

Louisiana's coastal lowlands are a mixture of marshes, swamps, natural levees, *cheniers*, barrier islands, open bays and other waterbodies. It is a physiographic province distinguished by the absence of relief. Mississippi River floods built the alluvial lowlands. For centuries, sediment-laden water fanned out along the coast creating two environments: a deltaic and a chenier plain (Shepard, 1971) (Figure 2).

East of Vermilion Bay, the deltaic plain contains the Mississippi's deltaic lobes formed during the last 6000–7000 years. Accumulation of material in this zone is a direct result of sedimentary processes within a delta system. Except for the modern 'bird's foot' delta, each lobe advanced into the continental shelf's shallow waters and was distinguished by numerous distributaries. As each channel bifurcated, diverted sediments prograded the coast seaward, building new land. Through time, recurring channel changes created an intricate 'horse's tail' pattern of levee fingers extending into the marsh.

Deltaic plain plant communities are composed of *flotant* (Russell, 1942; Chabreck, 1972). This is a vegetative assemblage distinguished by oystergrass *(Spartina alterniflora)*, water hyacinth *(Eichhornia crassipes)*, wiregrass (also called marshhay cordgrass or *Spartina patens)*, and *canouche (Panicum hemitomon)*. Anchored in a relative thin, matted layer of decomposing vegetable debris, *flotant* is truly floating on water or supported by highly aqueous organic ooze (Russell, 1942) – a condition locally described as 'trembling prairie'. These grass-derived materials develop in place. They are not altered by alluvial deposits; they continually decay and accumulate as peats. Decomposition maintains an organic layer that thickens with subsidence to a depth of 3 m to 6 m (Russell, 1942; Kolb and Van Lopik, 1958). Unless there is continuous sediment renourishment by flooding, land compacts, subsides, and sinks below the water surface (Nummedal, 1982).

West of the Mississippi's postglacial deltaic sequences, the chenier plain has a relatively smooth and regular shoreline. It is occasionally fronted by mudflats and

Figure 1. Louisiana's alluvial wetlands are a major link in the migratory pattern of numerous wintering waterfowl. Millions of birds utilize the wetlands in their yearly journey from northern latitudes to their winter homes in South America. (Photography courtesy of Louisiana's Department of Wildlife and Fisheries).

backed by marsh with an intervening series of beach ridges (Wells and Kemp, 1982). *Chenier* is derived from the French *chene*. It refers to the live-oaks (*Quercus virginiana*) that frequently grow on these ridges (Howe et al., 1935). In many instances, these topographic elements will be 1.5 m or more above the marsh. The sand and clay *cheniers* serve as land corridors (Figure 3), and nodal points, for rows of farms and hunting camps that follow their crests.

Each *chenier* marks a once active shoreline (Shou, 1967). When the Mississippi occupied one of its western courses, clays, muds and sands were carried westward by littoral currents advancing the chenier plain as a mud coast. Progradation interruptions permitted coarser particles to accumulate as a ridge. Increased sedimentation allowed the shoreline to advance, leaving the conspicuous, oak-covered *chenier* as an impressive and continuous feature (Howe et al. 1935). Silts and clays collected in the littoral zone and were stabilized by salt-tolerant vegetation creating a firm foundation.

Swamps and marshes

Swamps exist where soils are saturated for one or more months of the growing season, with water not too deep to prevent germination or to drown year-old plants. Along the Gulf Coast, swamp forests are best developed as part of river floodplains. Two plant species, bald cypress (*Taxodium distichum*) and tupelo gum (*Nyssa aquatica*) prevail.

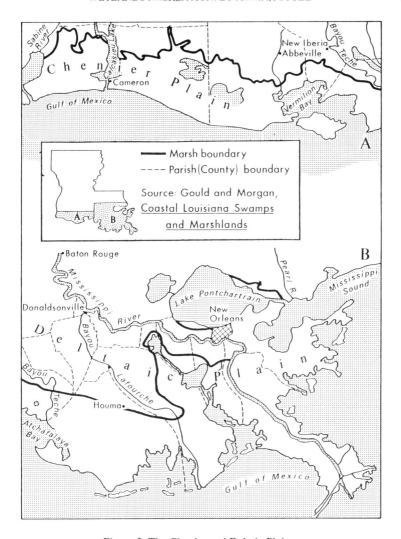

Figure 2. The Chenier and Deltaic Plain.

From a geological perspective, marshes are short-lived features. They are a product of currents, waves or rivers carrying sediments into shallow coastal waters. Through time, these deposits vertically accrete to a point where their surfaces can be vegetated by grasses that can tolerate tidal flooding and saltwater inundation. Once stabilized, they change water surfaces into a 'sea of grass'.

Marsh plants are classified as saline, brackish, intermediate and fresh and parallel the coast in an east-west direction (Penfound and Hathaway, 1938). The 20 ppt (parts per thousand) isohaline limits salt marsh. Oystergrass is the prevalent plant. Its abundance is in direct proportion to water table and salinity conditions. If these factors are changed, land surfaces assume aspects similar to a brackish or intermediate marsh. Brackish plants tolerate salinities of 10 ppt to 20 ppt. Marshhay cordgrass is the major species. A heterogeneous vegetative assemblage

Figure 3. Extensive cattle herds roam the Chenier Plain's marshes. In late fall cowboys move the livestock into the marsh and herd them out of the marsh in the early spring. (Photography courtesy of Louisiana's Department of Wildlife and Fisheries).

characterizes the intermediate band. These plants are capable of surviving in salinities that range from 2 ppt to 10 ppt. Salt grass (*Distichlis spicata*), roseau (*Phragmites australious*), bulltongue (*Sagittaria falcata*) and the far-ranging wiregrass are prevalent. Inland, freshwater marshes tolerate salinities of 1 ppt to 2ppt (Penfound and Hathaway, 1938; Wright *et al.*, 1960).

This aquatic habitat provides outdoorsmen with year-round recreation opportunities. In winter hunters, trappers, and fishermen harvest ducks, fur- and hide-bearing animals and numerous fresh- and saltwater fish. At this time, dead and decaying vegetation characterize the marsh. Seeds often blanket water surfaces, providing a valuable duck food.

In spring, water levels recede so turtles, and alligators (*Alligator mississippiensis*) come out of hibernation and various aquatic plants begin to germinate. This is the time for shrimp, crab, crawfish (Figure 4) (*Procambarus clarkii* and *P. A. acutus*) and fish. Firearms are replaced by fishing rods, reels and tackle. Until the first cold fronts move through the area, fishing and boating prevail. By late September, hunting season is open.

The recreationalist

South Louisiana's climate provides the recreational enthusiast with nearly ideal weather conditions. Warm summers and mild winters are the generalized weather conditions. Since rain falls all year, there is no dry period. In spite of nearly ideal

Figure 4. Crawfishing is a major recreation/commercial activity in the swamps. This is particularly true in the early spring. (Photography courtesy of Louisiana's Department of Wildlife and Fisheries).

weather and lengthy coastline, beach-oriented recreation is limited. There are few accessible beaches within the coastal zone. For the most part, Louisiana's outdoorsmen prefer to build 'camps' that are used as away-from-home bases for hunting, boating, and fishing.

Camps

There are more than 10 000 wetland camp (Gary and Davis, 1978). These recreational dwellings come in all shapes, sizes, styles and designs. Whether built on pilings above marsh mucks, or on a slab anchored to a natural levee, *chenier* or beach, the dwellings include roughewn, tar-paper hovels, piling-supported mobile homes suspended 4.5 m above the ground, simple bungalows, elaborate second homes, and a multitude of other architectural gems. On account of their dilapidated status, many are described as 'Tilten Hiltons'. Swimming pools, glass-enclosed verandas, and enough rooms and beds to accommodate 50 guests comfortably are normal for luxurious camps. Some two-storied units are equipped with outside elevators. Others are one-room 'houses' built for only the most rugged, and indefatigable individual.

Figure 5. This piling-supported camp represents one example of the types of structures found in the wetlands and is accessible only by water. (Photography courtesy of Louisiana's Department of Wildlife and Fisheries).

Figure 6. Hurricane Camille, a category 5 storm, did major damage to the fishing camps along the north shore of Lake Pontchartrain. (Photography courtesy of the United States Army Corp of Engineers, New Orleans).

Camps accessible by boats, are often equipped with generators, butane, and ingenious cisterns to make them a bit more comfortable. Those with highway access, are better built. Public utilities are often available (Figure 5).

Whatever design and costs are involved, camps are constructed to meet their owner's leisure-time needs: all can be considered multifunctional and seasonal. Based on utilitarian function, two types emerge: hunting and fishing (Figure 6). Hunting camps are utilized mainly during the fall and winter hunting season. They tend to be ramshakle buildings located in fresh and brackish marshes; sites preferred by wintering waterfowl. In contrast, fishing camps are used year-round. They are more elaborate and are located throughout the marsh, as game fish and certain shellfish are found in almost all wetlands habitats. In addition, several hundred barge-mounted camps can be towed to saltwater angling sites during summer and moved inland for hunting purposes.

Principal recreational activities

Hunting

Winter recreation endeavors focus on hunting. At least 100 000 individuals hunt within the marsh. These sport enthusiasts hunt waterfowl, puddle ducks, marsh deer (*Odocoileus virginianus*) and rabbits (*Sylvilagus aquaticus*). Demand for hunting space is so acute that two recurring costs are inescapable: a land lease or a membership in a private hunting club.

There are approximately 50 major marsh land owners. In order to hunt the marsh successfully, a lease must be obtained from one of these individuals or corporations. Cost varies, depending on location, size, number of ponds, and the tract's hunting potential. As a standard rule, better the hunting, greater the fee.

At the turn of the century, hunting clubs were organized to take advantage of their leased land. In some cases, these clubs control extensive hunting tracts. Colorful placards identify camps as 'Tropical Gardens Gun Club,' 'The Scrimp 'N Scrounge Duck Club', 'Carlton's Folly,' or 'Florence Hunting Club.' Louisiana's French heritage is reflected in camps called *C'est notre Plaisir, Chateau de Bateau, Lagniappe*, and *C'est La Vie*. Whatever whimsical name is employed, these dwellings are staging points for recreational activities.

One club owns hunting rights to 16 000 ha. Other clubs control smaller tracts but guaranteed hunting space is thereby supplied. On club property, facilities range from spartan to spacious. Some club members pay an initial fee from $1000 to $2000 and yearly dues of as much as $500. In such surroundings the only thing a member does is pull the trigger. All arrangements are made to insure a successful hunt. Often club employees kill the game, so members do not need to leave the clubhouse.

For those who cannot afford a lease or club membership, state operated wildlife management areas are free (Table 1). Within these preserves, hunting is controlled in an attempt to assure a never-ending supply of game resources. These management areas protect numerous avian and aquatic species. On wildlife refuges hunting is not permitted, as continued enjoyment of the outdoors depends on maintaining a healthy resource (Duffy and Hoffpaeur, 1966). Refuges are particularly important as nesting sites for recreationally important species. As one link in the North American Flyway, they help alleviate 'heavy shooting pressure' on the waterfowl population. During peak years it is estimated between 6 to 8 million birds will rest in Louisiana's marshlands (Burts and Carpenter, 1975).

Table 1. Wildlife management areas and refuges.

Wildlife Management Areas	Refuges
Atchafalaya Delta	Lacassine Waterfowl Refuge
Attakapas Island	Paul J. Rainey Wildlife Sanctuary
Biloxi	Rockefeller Refuge
Bohemia	Sabine Waterfowl Refuge
Manchac	State Game Refuge
Pass A Loutre	
Pearl River	
Point Au Chien	
Salvador	
Wisner	

Source: Brunett and Wills, 1978.

Louisiana's marshes, swamps and waterbodies constitute one of the nation's most important waterfowl wintering habitats (Table 2). Millions of birds rest and feed in the coastal zone before flying to their wintering grounds in Central and South America (Figure 7). During their winter stay, the migratory populations seem to prefer fresh to intermediate marshes, although some species can tolerate brackish conditions. They make hunting an extremely important and popular pastime. Managing this resource is accomplished by regulating hunting pressure within breeding, migration, and wintering areas (Herring, 1976).

Table 2. Louisiana's wintering waterfowl.

American Widgeon	(*A. americana*)
Black	(*A. rubripes*)
Canada Goose	(*Branta canadensis*)
Canvasback	(*A. valisineria*)
Gadwalls	(*A. strepers*)
Hooded Mergansers	(*Lophodytes cucullatus*)
Lesser Scaup	(*Aythya affinis*, known as a *dos gris*)
Mallard	(*A. platyrhynchos*)
Ringneck	(*Aythya collaris*)
Ruddy	(*Oxyura jamaicensis*)
Scaup	(*A. affinis* and *A. marila*)
Shoveler	(*A. clypeata*)
Snow Goose	(*Chen caerulescens*)
Teal	(*A. crecca* and *A. discors*)
Wood Duck	(*Aix sponsa*)
White-fronted Goose	(*Anser albifrons*)

Figure 7. To successfully hunt in the marshes or swamps requires a well trained 'bird' dog. (Photography courtesy of Louisiana's Department of Wildlife and Fisheries).

It should be emphasized that marsh recreational days may be dual in nature; hunting and fishing are often carried on during the same day. Consequently, recreational endeavors totaling at least 10 000 000 man-days (1 man-day = 24 hours) annually take place in the waterlogged lowlands (Davis and Detro 1975).

Fishing

Due to the large variety of fresh- and saltwater species, fishing is the largest recreational sport. Spotted trout (called a Speckled Seatrout or 'Specs') are the most sought-after saltwater species. Found in shallow waters, 'specs' average from 0.45 kg to more than 2 kg. In freshwater marshes, the quarry is a 2.2 kg or better largemouth bass. Even though at times they coexist with saltwater species, they prefer freshwater. Bass are difficult to catch and it takes 'bass savy' to land them consistently.

While not caught in the marsh, pompano is a prized game fish – a gourmet item considered by many as the best eating of all fish taken off Louisiana's coast. Until hydrocarbon exploration offshore, pompano, king mackerel, Atlantic croaker, jack crevalle and red snapper were rarely caught. Offshore production platforms became, in effect, artificial reefs. As new platforms were added, they enhanced the saltwater species feeding ground. With time, these 'reefs' became one of the world's finest offshore fishing provinces (Ditton and Auyong, 1984; Reggio, 1987; Wilson, n.d.).

The estuarine-marsh-swamp system is also a nursery ground that guarantees an abundant seafood supply. More than 90% of the Gulf's finfish spend part of their life-cycle within the coastal zone. Nationally, Louisiana's fisheries rank first in

Table 3. The principal sport fish of the coastal zone (After Jackson and Timmer, 1975).

Freshwater	
Bluegill	*Lepomis macrochirus*
Crappie (White and Black)	*Promoxis ssp.*
Largemouth Bass	*Micropterus salmoides*
Spotted Bass	*Micropterus punctulatus*
Striped Bass	*Morone saxatilis*
Warmouth	*Lepomis gulosus*
White Bass	*Morone saxatilis*
Yellow Bass	*Morone mississippiensis*
Saltwater	
Amberjack	*Seriola dumerili*
Atlantic Croaker	*Micropogonias undulatus*
Black Drum	*Pogonias cromis*
Bluefish	*Pomatomus saltatrix*
Blue Marlin	*Makaira nigricans*
Cobia	*Rachycentron canadum*
Dolphin	*Coryphaena hippurus*
Florida Pompano	*Trachinotus carolinus*
Jack Crevalle	*Caranx hippos*
King Mackerel	*Scomberomorus cavalla*
Little Tuna	*Euthynnus alletteratus*
Redfish	*Sciaenops ocellatus*
Red Snapper	*Lutjanus blackfordi*
Sailfish	*Istiophorus platypterus*
Spanish Mackerel	*Scomberomorus maculatus*
Sheepshead	*Archosargus probatocephalus*
Southern Flounder	*Paralichthys lethostigma*
Spotted Sea Trout	*Cynoscion nebulosus*
Tarpon	*Megalops atlanticus*
Tripletail	*Lobotes surinamensis*
Tuna	*Thunnus spp.*
Wahoo	*Acanthocybium solanderi*
White Marlin	*Tetrapturus albidus*

tonnage and second in value. Although commercially exploited, recreational enthusiasts contribute more than $400 million to the local economy, with more than one half million people involved in this leisure-time activity.

Sport fishing is a year-round enterprise. It changes with the various species' breeding cycles and summer months are preferred. At times, so many people seek marsh access that 100 or more automobiles may be seen at launch site.

To meet public demand for launching facilities, boat ramps dot the landscape; there are 150 to 200 launching ramps available for public use; some can accommodate up to 100 cars and trailers. As might be expected, they range from small, roadside ramps, able to accommodate 5 to 10 automobiles, up to private marinas with as many as four launching ramps or hoists. At many of these facilities, charter boats take small groups of fishermen offshore to troll or bottom fish for larger species.

Figure 8. Although shrimping is primarily a commercial enterprise, with many large vessels lining south Louisiana's bayous, recreational sportsmen also harvest this resource.

Once their boats are in the water, the outdoorsman can go after either fresh or saltwater species. Travel is easy, as the region is laced by an intricate system of natural and artificial waterways. These routes supply the connectivity necessary to exploit marshland resources.

Besides pole fishing two additional activities are important: crabbing and shrimping.

Crabbing

Crabbing is a family-oriented pastime concentrated in the summer. Blue crabs *(Callinectes sapidus)* are migrants. They stay in shallow waters in warm months

Figure 9. Louisiana's wetlands are eroding at an average rate of approximately 1 ha. every 45 minutes.

and move to deeper waters in winter. Recreational fishermen crab along roads bordering coastal bayous and drainage canals and in estuarine waterbodies. Trotlines and drop nets are traditional methods employed to capture this crustacean. The recreational catch exceeds the commercial catch by almost four times (Adkins, 1972) providing many families with a delightful seafood at nominal expense.

Blue crabs occupy almost all available habitats, with a large number caught in periods of low salinity and temperature. In warm weather, with abundant available food, a crab can mature in about twelve months. In attaining adult status each crab sheds its shell approximately 15 times. For two hours it becomes a soft-shell crab, considered a delicacy (Adkins, 1972). After two hours the crab is considered a 'paper shell'.

Shrimping

Prior to introduction of otter or shrimp trawls, Louisiana's shrimp catch was taken by seins. Cast nets are history; wooden and steel hulled boats are now outfitted to trawl coastal waters for white and Brazil (brown) shrimp (*Penaeus setiferus* and *P. aztecus*). In May, boats catch Brazilian shrimp (Figure 8). White are caught in the August to December season. A special license is now required to sport shrimp. These individuals can catch up to 45 kg day; a shrimping license is mandatory, and the catch cannot be sold.

Like their commercial counterparts, recreational shrimpers must trawl within the regulated season. To improve their harvest, some built large, winged 'butterfly' nets. These wharf-mounted *poupiers*, when lowered into a bayou, are an efficient way to 'trap' migrating shrimp.

Erosion, accretion and recreation

Sea level is rising. Over the last century a change of 15 cm was detected. Within 50 years the numbers may be between 30 cm to 60 cm (Begley and Cohn, 1986; Gilbert, 1986). Even a small shift upward will inundate Louisiana's coast. Concern over sea level rise has prompted publication of numerous projections. Most of these estimates are tied to carbon dioxide and other gases that effect air temperature, projected Polar Ice melt, and thermal expansion of the world's oceans. Land just above sea level may, by 2000 A.D., be flooded (Maranto, 1986; Titus et al., 1987; Walker et al., 1987). Tidal records along Louisiana's coast indicate 'relative sea level rise' is 9 mm to 13 mm yr^{-1}, or about 91 cm to 1.2 m $century^{-1}$ (Louisiana Geological Survey, 1987). This includes subsidence.

Property owners appear to downplay the threat. Nevertheless, much valuable real estate will vanish. As the sea rises, it will erode two to five times faster than it does today (Begley and Cohn, 1986). In 1980 Louisiana's marsh losses exceeded 10 000 ha. Since 1900, 450 000 ha were lost (Table 4). This is more land loss than all other United States' coastal states combined. Louisiana's land/water interface is under attack. The marshes importance as a recreational resource is also threatened (Figure 9).

Now confined within a conduit of artificial levees, the Mississippi River can no longer inundate its historic floodplain. Sediments are channeled off the continental shelf. Their loss deprives the lowlands of the 'material' necessary to build new land at a rate greater than subsidence. Nothing is available to offset the loss. Erosion is the end result. More than 100 km^2 of land disappear each year (Wicker et al., 1980;

Table 4. Rates of wetland loss (After Houck, 1983).

Year	Rate of Loss
1913	17.4 $km^2 yr^{-1}$
1946	40.9 $km^2 yr^{-1}$
1967	72.8 $km^2 yr^{-1}$
1980	102.1 $km^2 yr^{-1}$
1985	ca. 129.5 $km^2 yr^{-1}$

Gagliano *et al.*, 1981). When these near sea level surfaces are gone, the largest marshland in America will be a memory. Only saltwater recreational pursuits can be enjoyed.

References

Adkins, G. 1972. A study of the blue crab fishery in Louisiana. New Orleans: Louisiana Wild Life and Fisheries Commission, Technical Bulletin No. 3.
Begley, S. and Cohn, B. 1986. The silent summer. Newsweek, (June 23), 64–66.
Brunett, L. and Wills, W. 1985. A guide to wildlife management areas. Baton Rouge: Louisiana Department of Wildlife and Fisheries.
Burts, H. M. and Carpenter, C. W. 1975. A guide to huting in Louisiana, the hunter's paradise. New Orleans: Louisiana Wild Life and Fisheries Commission.
Chabreck, R. H. 1972. Vegetation, water and soil characteristics of the Louisiana coastal region. Baton Rouge: Louisiana State University, Agricultural Experiment Station. Bulletin 664.
Davis, D. W. and Detro, R. A. 1975. Louisiana's marsh as a recreational resource. *in* H. W. Walker (ed.), Coastal Resources, Vol. 12. Baton Rouge: Louisiana State University, School of Geoscience, pp. 91–98.
Ditton, R. B. and Auyong, J. 1984. Fishing offshore platforms, central Gulf of Mexico: An analysis of recreational and commercial fishing use at 164 major offshore petroleum structures. 1984. Washington, D.C.: Minerals Management Service, Gulf of Mexico Regional Office. OCS Monograph MMS 84–0006.
Duffy, M. and Hoffpaeur, C. 1966. History of waterfowl management. *Louisiana Conservationist* 18 (7, 8), 6–11.
Gagliano, S. M., Meyer-Arendt, K. J., and Wicker, K. M. 1981. Land loss in the Mississippi River deltaic plain. Transactions Gulf Coast *Association Geological Societies* 31, 295–300.
Gary, D. L. and Davis, D. W. 1978. Mansions on the Marsh. *Louisiana Conservationist* 30 (3), 10–13.
Gilbert, S., 1986. America washing away. *Science Digest* 94(8), 28–35, 75–78.
Gould, H. R. and Morgan, J. P., 1962. Coastal Louisiana swamps and marshlands. In E.H. Rainwater and R. P. Zingula, (eds). *Geology of the Gulf coast and central Texas and guide book of excursions*. Houston, Texas, Annual Meeting Geological Society of America, 287–341.
Herring, J. L. 1976. Sound reasons for hunting seasons. *Louisiana Conservationist* 26(11, 12), 4–8.
Houck, O. A. 1983. Land loss in coastal Louisiana: Causes, consequences and remedies. *Tulane Law Review* 58(1), 1–68.
Howe, H. V., Russell, R. J., McGuirt, J. H., Craft, B. C. and Stephenson, M. B. 1935. Reports on the geology of Cameron and Vermilion parishes. Geological Bulletin No. 6. New Orleans: Department of Conservation, Louisiana Geological Survey.
Jackson, P. M. and Timmer, D., Jr. 1975. A guide to fishing in Louisiana: The sportsman's paradise. Baton Rouge: Louisiana Department of Wildlife and Fisheries.
Kolb, C. B. and Van Lopik, J. R. 1958. Geology of the Mississippi river deltaic plain, southeastern Louisiana, Vol. 1. Technical Report No. 3–483. Vicksburg, Mississippi: United States Army Engineer Waterways Experiment Station.
Louisiana Geological Survey. 1987. Saving Louisiana's coastal wetlands: The need for a long-term plan of action. EPA–230–02–87–026. Washington, D.C.: United States Environmental Protection Agency.
Maranto, G. 1986. Are we close to the roads end? *Discovery* 7(1), 28–50.

Nummedal, D. 1982. Future sea-level changes along the Louisiana coast. In D. F. Boesch, (ed.), Proceedings of the conference on coastal erosion and wetland modification in Louisiana: Causes, consequences and options. Washington, D.C.: National Coastal Ecosystems Team, United States Fish and Wildlife Service, Office of Biological Services. FWS/OBS-82/59, pp. 164–176.

Penfound, W. T. and Hathaway, E. S. 1938. Plant communities in the marshland of southeastern Louisiana. *Ecological Monographs* **8**, 1–56. 2(1), 75–86.

Reggio, V. C., Jr. 1987. Rigs-to-reefs: The use of obsolete petroleum structures as artificial reefs. Washington, D.C. United States Department of the Interior, Minerals Management Service, Gulf of Mexico OCS Regional Office. OCS Report MMS 87–0015.

Russell, R. J. 1942. Flotant. *The Geographical Review* **32**(1), 74–98, p. 79.

Shepard, F. P. 1971. Our changing coastlines. New York: McGraw-Hill Publishing Company.

Shou, A. 1962. Pecan Island, a chenier ridge in the Mississippi delta plain. *Geografiska Annaler* **49A** (2–4), 321–326.

Titus, J. G., Kuo, C. Y., Gibbs, M. J., Laroche, T. B., Webb, M. K., and Waddell, J. O. 1987. Greenhouse effect, sea level rise, and coastal drainage systems. *Journal of Water Resources Planning and Management* **113**(2), 216–227.

Walker, H. J., Coleman, J. M., Roberts, H. H., and Tye, R. S. 1987. Wetland loss in Louisiana. *Geografiska Annaler* **69A**(1), 189–200.

Wells, J. T. and Kemp, G. P. 1982. Mudflat and marsh progradation along Louisiana's chenier plain: A natural reversal in coastal erosion. In D. F. Boesch (ed.), Proceedings of the conference on coastal erosion and wetland modification in Louisiana: Causes, consequences and options. Washington, D.C.: National Coastal Ecosystems Team, United States Fish and Wildlife Service, Office of Biological Services. FWS/OBS-82/59, pp. 39–51.

Wicker, K., DeRouen, M., O'Connor, D., Roberts, E., and Watson, J. 1980. Environmental characterization of Terrebonne Parish: 1955–1978. Final report. For the Terrebonne Parish Police Jury. Baton Rouge, Louisiana: Coastal Environments, Inc.

Wilson, C. A., Van Sickle, V., and Pope, D., no date. The Louisiana artificial reef plan: Executive summary. Draft report. Baton Rouge: Louisiana Geological Survey.

Wright, R. L., Sperry, J. J., and Huss, D. L. 1960. Vegetation type mapping studies of the marshes of southeastern Louisiana. College Station: Texas A and M University Foundation. Project No. 191.

Donald W. Davis
Department of Earth Sciences,
Nicholls State University,
Thibodaux, Louisiana, U.S.A.

14. The natural features of the Caspian Sea western coasts in the context of their prospective recreational use

Introduction

It is common knowledge that seacoasts play a major role in the most diverse domains of human activity. Their recreational resources rank prominently among the various natural resources that man can put to his use in coastal areas.

In the Soviet Union people usually go for seaside vacations and medical treatment to the Black Sea coastal areas, such as the Crimean southern and western seashores, and the Caucasian coast to the northwest, and to the Sea of Asov shores where natural features are of the same quality as those of the Black Sea. The Baltic Sea resorts are also very popular, even though bad weather spells are more frequent there and seawater temperatures are not high for the most of a high season. But the area's gorgeous nature, pine-tree woods and sea air combine to induce great numbers of holiday-makers to flock there.

In summer these traditional sea resorts are overcrowded with vacationers. For example, summer-time population of the resort town of Sochi swells to over one million people. The Crimean coasts also are crammed with hundreds of thousands or even millions of vacationers every year. Thus, today the Soviet Union's European part experiences drastic shortages of recreational territory with some of its most popular areas crammed with vacationers to a bursting point, on the one hand, while, on the other, vast parcels of almost virgin coastal area are available for development on the Crimean northwestern shore, Kerch Peninsula north and Caspian Sea western and eastern shores. It is for this reason that in recent years it has been proposed that the Caspian Sea shores should be developed as a national priority for recreational use, starting from its western coast.

The Caspian Sea western coast (Figure 1) extends entirely within the temperate warm semiarid zone. Its southern-most tip (Lenkoran coast) falls within the humid subtropics. The summertime seawater temperature rises to 25 °–26 °C, and in terms of the number of sunny days the Caspian western coast compares favourably with the Black Sea coastal area while its beaches are far superior in length to those on the Black Sea. The Caspian western coast is well developed and accessible to vacationers travelling from anywhere in the Soviet Union, for it has railroad, highway and air links to almost every center of the country. These factors allow to foresee a vigorous development of seaside vacationing facilities.

The natural features of the Caspian western coast

The climatic features of the Caspian western coastal area are controlled by its geographical position, meridian orientation, the type of circulation of air masses

Figure 1. Location map of the western Caspian Sea coast.

and adjacent relief. The climate is generally warm with a large number of sunny days. Solar hours average about 2000 yr^{-1}, as many as in Sochi. The city of Makhachkala air temperature averages +11.8 °C annually; it is +12.5 °C in Derbent and rises to +14.5 °C to the south. The warmest months are July and August, when the highest air temperature reaches 24–25 °C in the northern part of the coast (within Dagestan) and 25–27 °C in the south (Azerbaijan). The Azerbaijani coast bathing season lasts 5 to 6 months. The Dagestani coast suffers from regular shortages of rain: annual precipitation amounts to 250–450 mm, reaching the highest level in autumn (126–140 mm). The Kobystani coastal area (south of the Apsheron Peninsula) is also arid with precipitation never exceeding 200 mm yr^{-1}. But Lenkoran's annual precipitation record amounts to 1200 mm.

What is typical of the Caspian western coast is that its air pressure and temperature fluctuations are minimal and that drastic weather changes are also few in number (except in wintertime). These features are highly appreciated by numerous vacationers. The recreational value of the Caspian western coast is also favoured by rare fogs, scarce cloudiness in the summer (never over 36% in Dagestan); many frost-free days (up to 300 yr^{-1}) and negligible snow falling: for example, 5 days a year are recorded in the town of Khachmas.

The wind pattern is of course strongly influenced by morphological conditions. In the north the coastal plain looks like a narrow corridor between the mountains and the sea. The Arctic airmasses pass through this corridor on their way to the Russian Plain and farther on to Transcaucasus, moving along the coast. Tropical airmasses also pass through the corridor when high pressures establish over Turkmenia and southern Kazakhstan. This is the reason why northwesterly, northerly and southeasterly winds prevail over maritime Dagestan with southeasterlies prevailing north of the town of Izberbash and northwesterlies south of the town. Northwesterlies and southeasterlies account for most of the storms gushing at 30 m s^{-1} and faster. On the average, 15 m s^{-1} winds blow for 85 days in Makhachkala and 26 days in Derbent, their annual rate averaging 5.3 and 3.2 m per second respectively. Thus, the southern part of the Dagestani coast is less wind-swept and more comfortable than the north for climate-based medical treatment.

In contrast with most of the Dagestani coast, northerlies dominate the northern Azerbaijani coast and the northern shore of the Apsheron Peninsula, while northeasterlies and southeasterlies prevail on the south of the Apsheron Peninsula. Storms gushing at over 10 m s^{-1} blow 30–32 times a year, on the average, with northerlies or northwesterlies prevailing. They often bring substantial drops in air temperature, thereby reducing nice weather and vacationing time. In southern Azerbaijan, the number of days with storm winds (blowing at over 15 m s^{-1}) decreases to 24 in Kobystan and to 10 in Lenkoran.

Winds control also the wave conditions, although swells are as important to the Caspian western coast as wind-induced waves. For example, frequency of swells reaches 24% in Makhachkala, 21% in Izberbash, 26% in Derbent and up to 60% in Lenkoran. In Makhachkala, frequency of prevailing wind-induced waves is from NE – 19, from E – 19, from SE – 30 and from NW – 18 (in percentage points), while in Derbent waves are predominant from E (21%), SE (26%) and NW (37%). The highest waves recorded are 4.5 m in Makhachkala and 6 m in Derbent. However, low waves (up to 0.75 m) are prevalent in the Derbent area, as a rule. Along the Azerbaijani coast the roughest sea has been observed in the area of the Apsheron Peninsula and Archipelago, where in the vicinity of the Zhiloy Island, for instance, northerlies of storm force often generate waves 5 to 6 m high. Off to the south, waves from NE and SE prevail.

The wind pattern is also responsible for the emergence of alongshore currents, but these fade away as soon as winds cease to blow. In addition, winds are responsible for positive and negative set-up, which phenomenon is crucial to the shoaly coast. The storm surge may cause the sea level to rise by 1 to 1.65 m, entailing the flooding of beaches and of large areas of the coastal plain.

Sea level changes are one of the major factors affecting the dynamics of the Caspian coasts and serving as a criterion for appraising their recreational value. After high sea level at −22 m in the early 19th century (Berg, 1949; Leontyev, 1951; Kazancheev, 1956), the Caspian Sea level fell to the −26 m mark, remaining unchanged until 1929. Over that long period the established types of the Caspian coasts were roughly as follows: the northern area was dominated by very shallow water with mostly marshy coasts fringed by thickets of reed, with most of the area including the Terek, Volga and Ural deltas. In the eastern side prevailed retreating coasts cut in Neogene rocks of the Mangyshlak, Ustyurt and Kenderli-Koyasan plateaus to the north of the Kara-Bogaz Strait, and prograding coasts to the south of Cheleken that used to be an island in those days. In the western coastal area prevailing coasts were of the prograding type with clear alongshore drifts within the middle portion of the Caspian Sea, while small-size erosion segments with rather long accumulative coastal beaches were intermittently spotted in the area of the Apsheron Peninsula and southward.

Longshore sediment drifts were the most prominent feature of the dynamics of the western seashores. In the northern portion of the coastline (northern Dagestan) they moved in a north-west direction under the impact of prevailing waves, while in the Derbent area they were replaced by onshore-offshore sediment moving (the approach angle of the wave resultant equalled almost 90°) and in the northern Azerbaijani coast they stretched southward from the Samur delta up to the Kilyazi Spit (Nevessky, 1956; Leontyev, 1961). Thereafter other factors entered the picture, like prevalent cross-drifting and short longshore migrations of beach material in various directions.

This picture broke down drastically as a result of quick drops in the sea level in 1929–1940. The sea level continued to fall more slowly till 1977, when it reached its lowest mark of −29.2 m. Substantial reshaping of the Caspian coast followed. Reduced depth in the nearshore area brought about more destructive wave action, resulting in erosion and intensive moving of bottom sediments toward the shore. The early established sediment drift pattern was disrupted, the onshore drifting and the massive accretion of beaches became prevalent. The effect was christened after Bruun (1962), even though it had been described earlier by V. Zenkovich (1948) and O. Leontyev (1949). Along with offshore reshaping provoked by the lowering sea level, engineering works along rivers drastically reduced the sediment supply to the shore zone.

In 1978 the Caspian shores began a new period of their reshaping. Confounding numerous forecasts, the Caspian Sea level started to rise and is still rising now. According to predictions available, sea level will be 1.5 m higher by 1991 against the 1977 value (Leontyev, 1988). The rising level will inevitably result in erosion of the beaches and an advancing sea. The rising of the Caspian Sea level coincided with drastic reductions in sediment supply caused by damming of the Volga, Sulak, Kura and Samur rivers, which were major sources of terrigenous drifting materials. This exacerbated the erosion of the shores, especially in the western coastal area where terrigenous sediments are crucial to the shore (compared to the eastern shore, where biogenic and chemogenic sediment supply is predominant).

Coasts

Dagestani coasts. The Dagestani coast is a relatively narrow strip of coastal lowland (1 to 12 km) separating the slopy Caucasian foothills from the sea. Along the entire lowland, particularly where it gets wider, one can see clearly Pleistocene terraces located as high as 200 m above sea level. Along the shoreline, except for some short segments where the coast is formed of bedrock, there stretches the 'New Caspian Terrace', formed of Holocene marine sediments spreading to –20 and –22 m and consisting of shell and terrigenous sands. Not long ago the sand along the entire length of the strip had been shaped into dunes under wind action, but over the last two decades vegetation has stabilized the dunes. Numerous dunes have been destroyed by sand excavations for construction purposes, or levelled off for civil engineering projects. Over the last few decades a large area of lumpy drift sand has been fixed by trees planted for this purpose between Makhachkala and the mouth of the Sulak river. The same has been done to fix sand in the vicinity of the town of Kaspijsk.

Prior to 1929 almost all of the coast had been affected by erosion. A small cliff (2 to 5 m in height) was dug out in the loose sediments of recent terraces and more seldom in bedrock basis of ancient Caspian terraces. Even after the Caspian sea level had lowered, in 1929–1940, erosion continued to affect some limited sites, while over most of the coast low accumulative terraces emerged in front of the degraded cliff (the 1929 and 1940 terraces) and were fringed by a sand and shell beach 70 to 150 m in width.

Thus, until recently the Dagestani coast has had a high recreational potential, for out of its total of 430 km sand and sandy shell beaches stretch over 250 km, making it perfectly fit for seaside recreation. But the recent rising of the sea level has changed the coast dynamics substantially, and in some situations, impaired the quality of recreation. Sandy coasts development under conditions of rising sea level depends primarily on offshore area gradients, which vary greatly (from 0.0001 to 0.01) in the Caspian Sea. Where the slope is minimal, the coastal lowland is flooded by seawater passively. No discernible changes have affected the morphology of the coastal relief, for wave action has almost no impact upon this type of coast due to its excessive shallowness. Owing to the rising sea level upthrusting, groundwater has contributed to increased swampiness in those areas. This type of coast is typical of the Caspian Sea northwestern sector.

Rising sea level has introduced even more substantial changes in the shape of the accumulative coasts, with a steeper slope of offshore area (in the order of 0.001). It also contributed to flooding of vast parcels of the coastal lowland and on the seaside, the large shallow-water (20–30 cm) lagoon is fringed by a narrow barrier that rapidly advances landward. With the gradient of the offshore area growing steeper (roughly up to 0.01), the barrier joins to the shore and continues to move over the coastal plain, often burying its shrubbery under sand. It is only natural that such coastal areas should have lost much of their recreational value. This type of coast is widespread in Dagestan and is the most prominent feature of the area south of Kaspijsk and Makhachkala.

On accumulative coasts, with steeply sloped offshore area (over 0.01), the rising sea level has activated erosion of recent and Holocene accumulative features, and cliffs of 1 to 6 m high started to develop. Now these cliffs are numerous along the root of the Agrakhan Spit north of Makhachkala, in the Kaspijsk and Derbent areas and near the mouth of the Samur river among other places. Few are the cases where cliffs slope down right to the sea. As a rule, they are formed at the storm upwash level with a narrow beach stretching in front of them. Despite of preservation of recreational value of such types of coast, there still exists the danger that onshore facilities may be washed away. For example, the town of Kaspijsk embankment is destroyed quickly; the structures of a few boarding houses for vacationers north of Makhachkala and the first line of Derbent houses threaten to collapse.

Along Dagestan coasts there are numerous ridges of submerged wave-cut platforms which used to protect the coasts against erosive wave action prior to the rising of the Caspian Sea level. As a result of recent transgression, those ridges are now submerged in deep-water areas and can no longer perform a blocking function. The intensive erosion of portions of the coast at the back of the ridges followed. This type of coast prevails especially in the Izberbash area and in some sites south of Makhachkala and Kaspijsk.

Therefore now the rising sea level has made it more difficult to exploit the Dagestani coasts huge recreational potential, produced some negative effects upon its condition and raised the need for protecting the coast and its facilities against wave action. At the same time, according to observations, no substantial amounts of loose material have been reported to move alongshore as a result of the reshaping of the offshore area under the influence of the rising sea level.

Azerbaijan coasts. The northern Azerbaijani coast originates in the delta of the Great Samur river, affected by intensive wave erosion for a long time (Leontyev, 1961). Since 1950 a strip of land of over 300 m in width has been washed away, a part of the unique subtropical Samur forest reserve being already affected by erosion. The major reason of this erosion is a significant reduction (at 25 to 50%) in the river solid discharge, the important source of sediment supply to the shore zone, after the building in 1961 of a by-pass canal to provide the city of Baku with fresh water. The recent rising of the sea level has exacerbated the erosion problem facing the Samur delta.

The wide and complicated Samur-Divichi lowland runs to the south. Its northern part (adjacent to the Samur river) is a vast alluvial and proluvial debris fan made up of sand and shingles and formed over almost the entire Pleistocene. Until 1929 the marine edge of the fan near the Samur delta had been intensively erosioned by the sea, and cliffs of 6 to 7 m in height had been formed. After the sea level dropped in 1929–1940, wide and low wave-built terraces emerged at the foot of the cliff, with erosion continuing only on its northernmost tip. Modern-day transgression resulted in the flooding of young coastal terraces and the renovation of the cliff erosion.

The southern part of the lowland is a terraced coastal plain made up of accumulative mid- and upper Pleistocene terraces. Right along the sea runs the sand and shingle Holocene New Caspian terrace visibly affected by wind-borne action. The bulk of the terrace is formed of a huge barrier separating a large lagoon from the sea. As early as the 1950s, mobile dunes were a typical feature here, forming the highest part of the barrier (up to 10 m above the Caspian Sea level), while west of the barrier there was the big Akhzibir Lake, a relict of ancient lagoon. By the onset of modern-day transgression, the dune-shaped landscape had already been replaced by gently sloping lumpy sands fixed by vegetation with just a few marshy spots, remnants of the former lagoon. The area grew swampier as the sea level began to rise, followed by the upwelling of the ground water table.

South of the town of Siazan the coastal plain narrows sharply and the foothills of the Great Caucasus extend close to the shore, their structures continuing on the shore and offshore. As a result, right within the coastal area and on the seabed, bedrocks outcrop (Cretaceous bedrock and Paleogene and Neogene bedrocks to the south). However, along this entire coastal strip (as far as the Kilyazi Spit) there used to be well-developed compacted sand beaches of 100 to 150 m in width. They were backed by mobile dunes almost everywhere. At present the coastal landscape has been considerably changed by the new hydrodynamic conditions. The dune relief has been either destroyed or completely fixed by vegetation for the reasons mentioned above (wetter climate, the rising level of underground waters and man economic activity). The coastal areas of modern low terraces are in part flooded or wetted, becoming reed-overgrown lagoon marshes that are separated by a young beach ridge from the sea. The coastal wetlands are absent in areas with steep slopes, and behind of the beach ridge small and more or less permanent lagoon-type bodies of water are formed. The beach ridge is usually made up of bottom sediments (shells or sandy shell deposits), serving as evidence that the upper part of the offshore zone is vigorously erosioned. The beach ridge has a clearly shaped asymmetrical profile and apparent traces of active advancing over the coastal plain: fresh tongues of storm overwash the vegetation and outcrops of compacted clays and growing shrubs emerge on the beach.

Shell material was intensively piling up 10 years ago right in front of the Kilyazi Spit, where shells were ejected from the seabed due to the waves approaching the shoreline almost perpendicular. Here the shore accreted fast by forming steep and high (up to 2 m) shell beach ridges. Today this coastal area is being visibly eroded at the rate of roughly $0.5-1$ m yr^{-1}.

The former size of northern Azerbaijan's sand and shell beaches can be evidenced by an accumulative segment located between the Kilyazi Spit and Apsheron Peninsula, two major bedrock protuberances. Even under changed hydrodynamic conditions these shoreline protuberances still maintain wide sand and shell beaches, although one can see some evidence that the coastal area is under reshaping: ephemeral, shallow-water lagoons are formed there to be followed by more permanent ones resulting from the waves washing over the present-day beach ridge.

South of the Apsheron Peninsula and within the Kobystani coast, there is a relatively wide plain, backed by the foothills of the Great Caucasus and a series of mud volcanoes with Old Caspian terraces on their slopes. There exist wide sand beaches that have not been heavily affected up to the present by the rising of sea level. But scarce population of this area, arid climate and shortage of fresh water combine to make the prospect of a recreational use a difficult perspective.

The Kura modern and ancient deltas appear as a flat plain, partly swampy and partly covered with mobile dunes and lumpy eolian sands. The dams that got under construction in the Kura valley as early as the 1960s have cut almost in half the river solid discharge, leading to shortages of drifting material in the shore zone. This affected, above all, the evolution of the river delta and its adjacent accumulative beach areas fed by river-supplied alluvial deposits. As a result, the marine edge of the delta and the Kura Spit expanding to the south began to be washed out tangibly; this erosion was increased by rising of sea level in the last years. Now the Kura Spit is washed out at the rate of roughly 15 m yr^{-1}. The dune cluster that used to extend along the spit is now destroyed by the sea and in 1985 waves broke through the spit root part, turning its southernmost tip into an island.

Despite heavy rains falling mostly in autumn and winter the Lenkoran coast compares favourably with the regions located farther to the north in terms of its natural features. Its subtropical climate, beautiful gravel and pebble beaches, thick woods that cover the slopes of the Talysh Mountain Range approaching the sea combine to render the coast very promising for future recreational use. But the modern-day rising of the sea level has increased erosion of the beaches and in some areas called for immediate action to protect some of the economic objects against destruction.

Conclusions

At present Azerbaijan and Dagestan accumulative coasts are crucial to the vigorous economic and recreational development of the Caspian Sea littoral but the drastic rising of the sea level has introduced substantial adjustments in the planning and implementation of proposed activities. Now that the Dagestani and Azerbaijani coastal natural features have been highly appraised overall as promising seaside recreational zones to be enjoyed by masses of people, close attention is to be paid to the complications resulting from the sea modern-day transgression. The rising sea level exacerbates, above all, the erosion of the beach in areas where it used to be washed out for different reasons (mainly as a result of man activities). What is more important is that the erosion is now spreading to the new areas where earlier the coast was stable or accreted. Meanwhile beach stability is crucial to recreational development projects. Sea erosion of the Caspian is a new phenomenon, for never over the last one hundred and fifty years has it been necessary to protect the shores agains wave erosion. Today this factor assumes a major importance for those planning various littoral development projects, notably for recreational purposes.

Rising of underground water tables is another negative after effect of present-day rising of the Caspian Sea level. It results in the swamping of the coastal strip, especially in low-coast areas that have a wide zone of present-day nearshore accumulation. Swamping encompasses the part of low terraces that adjoins a modern beach usually shaped as a full-profile beach or beach barrier. The bogged nearbarrier part of modern terraces is a feature prevailing in numerous areas of the Caspian Sea western coasts. This feature impedes access to the beach for vacationers, on the one hand, and on the other, obstructs the construction of even light-weight building structures necessary to service beach activities. Swamping is sure to complicate the laying down of various communications.

It is natural that an importance of wind surges is increased as sea level rises. In view of the fact that the sea level will continue to rise, at least, till 1991 and will, it appears, have reached the −27.5 m altitude as a minimum by that year, various capital construction projects for building recreational facilities are to be launched in areas that lie not lower than the −23 m, that is only within the Holocene New Caspian terrace.

References

Berg, L. S., 1949. Essays of physical geography. USSR Academia of Sciences, M., 280.
Bruun, P., 1962. Sea level rise as a cause of shore erosion. *J. Waterways and Harbours Div.* **88**, 117–130.
Kazancheev, E. N., 1956. New data about the Caspian Sea level at the end of XVIII century and in early XIX century. *Izvestia VGO* **88**, (6), 549–551.
Leontjev, O. K., 1949. Profile reshaping of accumulative coast due to sea level dropping. *DAN USSR* **66**, (3), 377–379.
Leontjev, O. K., 1951. The shoreline evolution of the northern Dagestan coast of the Caspian Sea. *Izvestia VGO* **75**, (4), 353–363.
Leontjev, O. K., 1961. Main features of the morphology and evolution of the northern Azerbaijan coast of the Caspian Sea. *Trudy Okeanograph. Komission* **8**, 3–32.
Leontjev, O. K., 1988. Problems of the Caspian Sea level and stability of Caspian shores. Vestnic MGV, ser. geogr., I, 14–21.
Leontjev, O. K. and Khalilov, A. I., 1965. Natural conditions of the Caspian Sea coasts formation. Azerb. Republ. Academia of Sciences, Baku, 206.
Nevessky, E. N., 1953. Sand sediment transport along the western coast of the middle Caspian Sea. *Trudy Inst. of Okeanol. of USSR Academia of Sciences* **7**, 154–159.
Zenkovich, V. P., 1948. Cutting of erosion profile under process of sea level rising. *DAN USSR* **63**, (2) 292–295.

O. K. Leontjev, S. A. Lukyanova and L. G. Nikiforov
Department of Geography,
Moscow State University,
U.S.S.R.

SECTION III

Planning for Recreation

15. Construction of a recreational beach using the original coastal morphology, Koege Bay, Denmark

Introduction

Every Dane is familiar with the coastal environment. No dwellings are situated more than 50 km from the nearest coastline. The population of Denmark amounts to a little more than 5 million, and with a coastline of 7314 km there is 1.5 km coast per inhabitant, which is considerable also by international standards.

Unfortunately this coastal resource is unevenly distributed over the country. The Metropolitan Region thus comprises 6% of the total area of Denmark (43 100 sq km) and 7% of the coastline for a population of 1.7 million corresponding to one third of the total. There is thus only 0.3 km coast available to each individual. But this is just an average figure and does not show the real demographic pressure on the coast. Coastal stretches north and southwest of Copenhagen are now densely built-up and with the growing population the need for reservation and development of coastal recreational areas has become more and more urgent.

North of Copenhagen, towards Elsinore (40 km), along most of the coastal area upper class houses prevail. Beaches for bathing and recreation for the city population are therefore restricted to the north coast of Zeeland or the Koege Bay (Figure 1).

As early as in 1936 the first project of creating a beach park just south of Copenhagen was elaborated. The designated area was situated only 15 km from the centre of Copenhagen 'a suitable area for citizens going there by bicycles'. Since then many project proposals saw the daylight (Figure 2) but it took 37 years before site investigations and design preparations were finally accepted. During this period the build-up of the southern part of Greater Copenhagen was explosive. Solely in the 1965–80 period 42 000 dwellings with the infrastructure required were built along the coast of Koege Bay.

In 1975 the Ministry of Environment decided to sponsor the project. The goal of the plan was: 'to establish a beach park facility, which at the same time preserved inherent environmental values and expanded the nature-given recreational resource for the benefit of the growing population'. Two other major objectives were provided by this project: to protect the hinterland from storm floods and to enlarge the harbour capacity in the bay for pleasure boats.

In April 1977 bulldozers and vessels for sand dredging invaded the area, and 2 years later the major part of the constructions were accomplished. In June 1980 the Prime Minister officially opened the 'Koege Bay Beach Park', including: 5000 ha (1200 acres) recreational area, 8 km of sandy bathing beach and 4 harbours with capacity up to 5000 pleasure boats. One hundred thousand visitors have been estimated on the new beach on a good hot summer day.

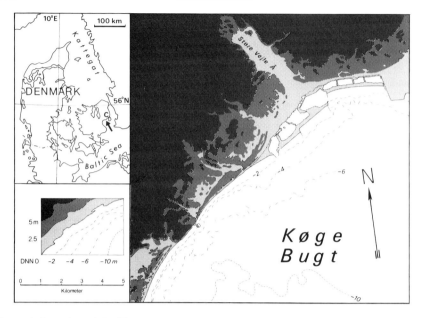

Figure 1. Location of the Koege Bay Beach Park. (C) Copenhagen. Notice the fossil barrier systems outlined by the 2.5 m contour.

The coastal environment

During the Quarternary period all of Denmark was covered by ice, and the initial landforms of Koege Bay were created during the last phase of glaciation. The bay can be characterized as a central depression reflecting a large glacier lobe, advancing in a northwest direction during the last stage of the Pleistocene age. When the ice melted away, a smoothed-out till plain was left, sloping gently towards the central part of the bay. Only in the northern part of the bay the monotonous landscape is disturbed by a glacial tunnel valley, St. Vejleá Dal (Figure 1). 6–9000 years ago, the base level was about 18 m below present sea level, and small rivers excavated the glacial valleys. The development of the actual coastal zone is due to the transgression during the Litorina Sea (5–6000 B.P.). The river valley was submerged and large quantities of sediment were deposited; simultaneously the waves affected a wide coastal zone because of the very small gradients. As the material of the sea bottom is rather rich in clay, stones and boulders, the creation of an equilibrium profile was slow. At the transgression maximum, about 5000 years ago, the relative sea level was 3 m higher than today. Onshore net transport of material predominated and therefore large sand masses, with an average grain size of 0.18 mm, accumulated along the coastline and formed several, extensive barrier- and lagoon systems at different levels (Figure 1). In 1972 the coastal landscape (Figure 3A) shows a single, sandy ridge – an initial barrier – in the nearshore zone deflecting the coastline because of a gradually decreasing gradient north of Mosede

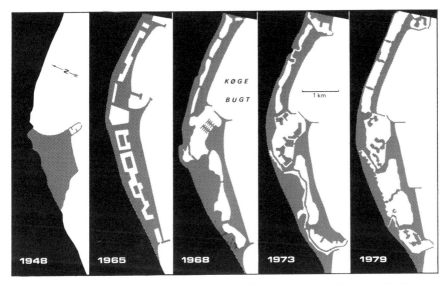

Figure 2. Earlier project designs of the Beach Park compared to the one realized.

(a small fishing port). The development of the nearshore off the Broendby Strand differs markedly and is characterized by multiple and minor bars. The change in the coastal development near the outlet of St. Vejleaa (river) can be explained by the different gradients of the nearshore zone. Surveyings of the surface beneath the marine deposits (VBI, 1965) show, that the valley of St. Vejleaa turns east in a more or less coast-parallel direction (notice the sea bottom contours on Figure 1). This means that the northern valley side supplies sediment to the nearshore and exposes a steeper gradient than the western part, which is dominated by the bottom moraine surface.

During the many years of discussion concerning the beach park, small temporary sand islands have been observed, and since the beginning of the 70s they have developed into more permanent barriers. The reason why the rate of accretion of material has been relatively slow is a consequence of the sea bottom geology. As indicated above, the sand fraction in the till is small and the great majority of the available and mobile sediment is already concentrated in the nearshore zone. Investigations of the sea bottom below –3 m (Nielsen 1984) reveal that the till is covered by a thin layer of lag material (coarse gravel and stones). Separation of the sand fraction from the till by waves and current is consequently a slow process.

According to Hayes (1979), Koege Bay must be classified as a low wave energy environment with $H_{bmax} < 1.5$ m. The principal reason is the orientation of the coastline in relation to the dominating westerly winds combined with the limited fetches. The astronomic tide is less than 0.1 m, but temporal wind tide may be pronounced. Sea level fluctuations of –0.8 to +1.1 m in 24 h are not uncommon and high waters near +2 m are recorded.

When evaluating the artificial impact on the natural coastal development it may be important to notice that no barriers in Koege Bay were present at the turn of the

Figure 3. Aerial view of the coast before and after the construction of Koege Bay Beach Park. A. May 1972. In the western part of the photo the nearshore is dominated by a broad and partly submarine ridge of sand enclosing a lagoon and salt marsh; on the light sandy area several narrow initial barrier islands can be detected. Towards the east the nearshore is characterized by a multiple bar system. The coastal stretch was here affected by waves (white brim). The asymmetric inlet is a consequence of an initial, now drowned river valley.

B. April 1985. 7 years after completion of the Beach Park. The inheritance of the original coastal geomorphology can easily be recognized.

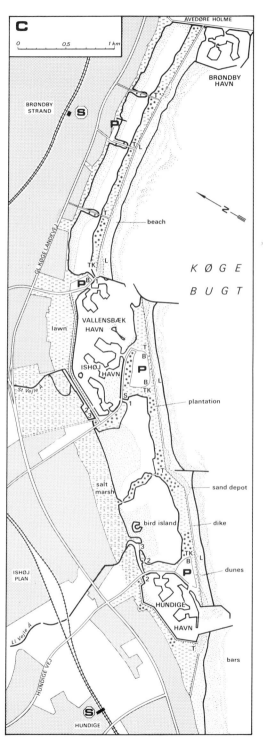

Figure 3. C. The main outline and functional organization of the Beach Park: (1) sluices leading sea water into the lagoons; (2) outlet sluices; (3) river water spill-overs; (K) kiosks and mini bars; (T) toilets; (B) bus terminals; (P) car parking; (L) life guards.

Terms of construction: 1976 tenders were invited to the coastal engineering works, and already in spring 1977 sand dredgers commenced the excavation of the lagoons. This material consisted of fine sand and was tubed to the new coastline. Simultaneously 3 175–m groins were built in order to keep the sand. During the winter 1977/78 coarse material was dredged from the sea bottom of Koege Bay and was placed as a covering layer to reduce marine and eolian sand transport. Bulldozers dispersed the sand pumped in according to the plan design, and finally the 'dunes' were put on the dike to convey some naturalness to the artificial morphology.

In all essential the Beach Park was accomplished at the turn of the year 1977/78, and during the following 1.5 years details behind the dike were completed. Only the harbour facilities still remained; this took another 1.5 years.

Total expenses (except for the harbours) in 1980 prices in US$: 20 million.

© Kort-og Matrikelstyrelsen, Denmark (A92–89).

century. Initial barriers appeared on maps for the first time in 1909, and since then they have slowly migrated towards the shore increasing their top levels. The origin of this barrier formation may be classified as proposed by Otvos (1980): barrier emergence through nearshore aggradation.

Creation of the beach park

The establishment of the Beach Park had the character of a land reclamation, by means of extensive beach nourishment. Sufficient loose materials were not found in the immediate vicinity, but sample investigations showed that the necessary amount and the right quality of sand could be quarried from the central part of the sea bottom of Koege Bay. To reduce the costs of transportation, the lagoon areas were extended. An extra groin was placed in the middle of the central beach stretch, which made it possible to establish a parallel displacement of the coastline. By this manipulation most of the dike length could be placed on the uppermost part of the initial barrier islands. The coastline adjustment further meant that the orientation of the coast would become perpendicular to the direction of the prevailing direction of the incident wave energy, – very important to minimize the longshore transport of sand.

In total, 5 million cubic metres of sand were needed for the alteration of the coastal profile. Three mill. of this amount were gained from excavation within the lagoon areas and from the harbour basins, and the last 2 million cubic metres were dredged from the sea bottom of Koege Bay.

Dike, dune and beach. The core of the artificial new coast is a 20 m wide dike, built of sand with a top level of 3 m DNN (Danish Ordnance Datum) (Figure 4). In dike sections, where particularly intense traffic was expected, for instance at the parking areas, 1–2 m high 'dunes' were created upon the crest of the dike (Figure 5).

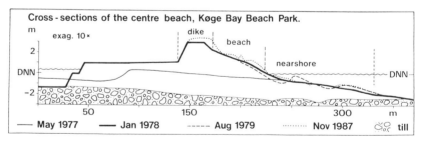

Figure 4. Cross-sections of the original and the artificial coast. The 1977-profile was surveyed just prior to the construction work and shows the morphology of an initial barrier (according to the form geometry, a giant nearshore bar is maybe a better term). Until now, the level of the inner nearshore is apparently affected slightly as compared to the shape given by the constructors. The matter is different concerning the beach. In front of the dike and on the wide backshore, large dune formations have developed. Notice the restricted amount of marine sediment superimposing the till.

Moreover, at the base of groins and harbour piers, dune-areas were carefully shaped to offer lee-giving pots to the benefit of the bathing guests. Besides the scenic values of the coastal landscape, these dunes are meant for emergency sand depots in case the sea threatens the dike system.

The beach in front of the dike was designed to have a width of 45 m with a slope from the foot of the dike (2.25 m DNN) to the coastline of 1:20 (Figure 4).

To subdue an undesired eolian sand transport, especially during the first years after termination of the constructions, relatively coarse sand (grain size: 0.4 mm) and gravel were required for the exposed layer of both dunes and beach.

The lagoons. To preserve the inherent environment, lagoons should still contain salt water after establishment of the Beach Park. To avoid stagnant water, automatic sluices were provided to force the water circulation in a one-way direction through the lagoons. The sea water is led from the central harbour area (Vallensbaek) into the lagoons both towards east and southwest (Figure 3C) during situations with higher water level in the sea than in the lagoons. Thus with water levels above +0.3 m DNN, the sluices will automatically close.

When the sea level drops below the level of the lagoon, the lagoon water escapes through sluices located at Hundige and Broendby Harbours. Recordings and calculations of the hydrography show that, with a daily sea level variation of 0.1 m, a total exchange of lagoon water would take place during a two-week period.

The depth of water in the original very shallow lagoons was increased in a zone parallel to the dike for the use of smaller boats like rowing boats, dinghies and wind surfers. In the three eastern lagoons, the bridges leading to the beach got an elevation of 4.4 m and a span of 12 m to allow boat-racing.

In many places the landward side of the lagoons is preserved as salt marsh areas and reed swamps. Consciously, these areas are made difficult to penetrate to secure quietness for the wildlife, especially the birds of passage which in great numbers rest at this locality (notice the 'bird island' on Figure 3C).

Before the construction of Koege Bay Beach Park extraordinary high waters during winter gales flooded large areas around the rivers St. Vejle Å and Ll. Vejle Å. It is tried to solve this problem by building dikes and sluices to regulate the river outlets (Figure 3C). The sluices are open, when the water in the sea is lower than that of the rivers, but they close when the water level exceeds +0.3 m DNN. To avoid inundation from river water in long periods with closed sluices, spillovers are established with top levels of +0.5 m DNN. During these events the lagoons are functioning as reservoirs.

Plantations. On the basis of botanical research in the whole coastal environment of Koege Bay, 20 species of trees and bushes were chosen for planting in the beach park; among others, oak, birch, European aspen, poplar and blackthorn, sea buckthorn and dog-rose. Immediately after the finishing of the construction works, the dike and 'dunes' were planted with lyme grass, and heavy watering and fertilization resulted in fairly natural-like dunes at the opening of the Beach Park to the Public (Figure 6).

Figure 5. Artificial dunes placed on top of the dike, (Photo: N. Nielsen, Oct. 1977).

Figure 6. Dike and 'dunes' shortly after plantation of lyme grass (Psamma arenaria). (Photo: N. Nielsen, March 1978).

Behind the dike, a 50 m brim of dune-roses was planted. No paths were established because visitors were expected to trample the most appropriate ways from the parking areas to the beach. Later on these paths were to be consolidated, but so far it has not been necessary.

Groups of trees are placed behind the dike too, for instance in connection with the parking areas and the service sites. The tree species vary from group to group, to avoid monotony in the landscape.

Geomorphological development after 1978

The creation of Koege Bay Beach Park has fixed a series of natural coastal landforms, which up to 1977 had still not been adjusted to a dynamic equilibrium. Since 1930 the salt marsh top level increased at a speed of 1–3 mm yr^{-1} in average and expanded into the lagoons (Frederiksen, 1972). Now the conditions for further growth have been reduced, but not ceased, as a consequence of reduced wave activity. Regarding the reed-swamps, cut-back is necessary now and then to keep a free water surface in the lagoons.

A pre-condition to preserve the original vegetation in the lagoons is a frequent addition of salt water. Here some problems have turned up, as the mean water level in the sea during the months of autumn is normally +0.3 m DNN, i.e. the level at which the sluices automatically close. Long periods with closed sluices have revealed that the river dikes are too low, and several times the water levels in the rivers have exceeded the spill-overs. In such situations the lagoons receive an undesired amount of fresh water.

The stability of the new coastline. Radical interference with natural coastal environment shows numerous examples of unforeseen ways of development, even if the chosen dimensions of nearshore, beach, and dunes have been decided from calculations based on recordings of the process agents in the affected area.

It was assumed that the initial barriers and bar systems would have developed into a more permanent barrier, such as seen at Jersie Strand in the southern part of Koege Bay, about 20 km from the Beach Park. The Jersie barrier was studied by Nielsen and Nielsen (1978), and the coastal processes and geomorphology are very similar to the area in question. These studies have shown that the barrier growth takes place stepwise. During low to normal wave conditions, sand depositions occur on the inner nearshore terrace landward of the breakpoint bars; nearshore bars are formed and migrate towards the foreshore of the barrier. During onshore storms associated with high water levels, the barrier may be overwashed. The total sand body of the barrier is displaced landward, while simultaneously increasing in height. This development continues until the gradients of beach- and nearshore morphology, counterbalance the waves each other within a certain sweep zone. By now, the Jersie barrier development has almost reached this stage and stabilized at a top level of 2–2.5 m DNN.

As mentioned, the sea dike was placed on an initial barrier which was not in an equilibrium position with the dynamic environment. The artificial beach has been established to a level (+2.25 m DNN) corresponding to a 'mature' barrier in Koege Bay. When the sand masses of dike and dunes are added, there is only a slight risk of a wave break through during onshore storms, so far at least.

An important question is, however: how will the level of the nearshore develop? In the first hand the gradient off the Beach Park looks steeper as compared with the profile of the natural barrier. Investigations at this coast after building of the Beach Park (Nielsen and Nielsen, 1984, 1985) and surveys in 1982 and 1987 indicate no retreat of the coastline, as might be expected (Figure 7). On the contrary, the nearshore terrace appears relatively stable, with an even positive trend in the material exchange. Concerning the beach, substantial amounts of sand are added as dune formations both in front of the dike and on the middle of the backshore.

The covering layer of coarse sand, gravel and rubbles, placed to strengthen the artificial dunes and beach to meet the eolian transport has now gone through a natural sorting process. The dunes on the beach are therefore 'real' and composed of homogeneous sand. The rest of the beach surface sediments have become markedly finer, and the stones do not bother bathing guests.

The landscape architects imagined an 8 km long and 45 m broad and white 'lido' (Figure 8). But this vision did not agree with the laws of coastal geomorphology in Koege Bay, where the normal beach width is approximately 10 m. To-day marine processes have adjusted the morphology of the beach to a broadness of 10–15 m bordering an almost continuous and lyme grass covered dune ridge (Figure 9).

The artificial dune. The dunes on top of the dike and in the sand depots now exhibit a good example of creating an intentional environment, when scientific knowledge cooperates with technological capability. When the dunes were made by man, they were an artificial product in every respect (cone-shaped hills of coarse materials), but to-day they look amazingly natural-like with their cover of eolian-sorted sand, caught by the dense carpet of lyme grasses. Moreover, the single dune forms have been modelled by the wind, from bulldozed heaps to more rounded accretional forms of the wind.

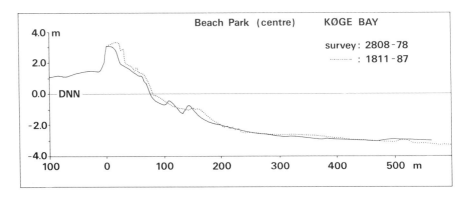

Figure 7. Surveys of a fixed cross section at the central part of the Beach Park. Except for the break point bar zone the morphology of the submerged profile seemed stabile. Accretions along the dike and on the backshore are pronounced.

Figure 8. The beach planned by the landscape architect, – a 'white lido' about 45 m broad. (Photo: N. Nielsen, 1978).

Figure 9. A 45 m wide beach does not agree with the dimensions of equilibrium beaches in Koege Bay. Dunes made by the wind now cover the whole dike front (left) and have spread over the man-made backshore. The actual width of the beach is now about 10 m (Photo: N. Nielsen, Nov. 1987).

The creation of dunes was a success, but skilful planning may have been combined with luck. For instance the designers may not have been aware of the fact that the prevailing winds blow from the west and therefore coincide with the orientation of the coast. This meant that most of the wind-borne material did not escape into the lagoons or the sea, but remained where it belongs to: the dune area.

Depositional problems. According to calculations, the longshore transport was not expected to cause any problems, as the coastline was created at a right angle to the prevailing wave energy. However, since 1977 a considerable accretion has taken place on a 3 km long coastal stretch southwest of the beach park. This development has been of great inconvenience to the shore-landowners.

For several years the level of the nearshore zone southwest of the beach park has been very high (Figure 3A), but after the creation of the Beach Park many barrier islands and lagoons have emerged, and in some sections the coastline is displaced more than 100 m in a seaward direction. The major disadvantage from this development is that sea weeds of different kinds are swashed into the lagoons during storm situations and infect the air when rotting in the stagnant water. Another complaint from the landowners is the reduced sea-view from their houses.

It is postulated that the Beach Park is the reason for the calamities, but preliminary interpretations of air photos, levellings and soundings in fixed profiles do not indicate longshore transport towards southwest. There are no stoss-side accumulations at the groins, and it is remarkable that during the same span of time the new beach has got a surplus of sand.

From a coastal geomorphological view-point the problem is interesting. Is the impressive beach growth just due to the fact that the Beach Park was built in the phase of barrier development when the barrier islands were getting a permanent character, or did the new coast in spite of all theoretical calculations accelerate this process anyhow? And what about the longshore transport which should be eliminated? Observations of the outline of the breakpoint bar passing the entrance to Hundige Harbour show that some sand movements do take place.

When studying the sea bottom off the nearshore there is a very distinct border between areas with and without mobile sediments. This line seems to move slowly shoreward with a simultaneous positive shoreline displacement. Assuming that the total volume of sand does not change essentially, the result will be a concentration of sediment and consequently a steeper profile. It is still not to be answered why this development is going on, but the problem is planned to be headlighted in a coastal research programme in 1988–89.

A source of sediment may originate from the deeper sea bottom. Air photo analysis and submarine investigations have shown that narrow (about 100 m wide) and thin (0.2–0.3 m) transverse sand ribbons are connected with the nearshore (Figure 3A), but, although their occurrence seems common all over the Baltic Sea and has been described by many coastal geomorphologists (concerning Koege Bay, Nielsen, 1984), the dynamic reason for their existence is still uncertain.

Final remarks

Apparently there are no menacing prospects in the development of the Beach Park's coastal geomorphology or along the adjoined coastal stretches. But of course it must be recognized that the construction is still only 10 years old, and that the general speed of development of this low wave-energy coast is slow.

It is also suggested that another and very important explanation to the success of this large-scale coastal interference is due to the fact that its design was based upon the character of the original geomorphology.

When evaluating this thorough change of coastal landscape it should finally be noticed that nature environmentalists only have had very few critical objections to the project.

Acknowledgements

Thanks are indeed due to my colleague, ass. prof. Jørgen Nielsen for collaboration during many years of field work in Koege Bay and for critical but helpful discussions. Kirsten Winther is thanked for improving the language, and John Jönsson for the final drawings.

References

Frederiksen, A. A., 1972: Report to DSB working group concerning airphoto interpretation of the Vallensbaek area. Geographical Institute, Copenhagen.
Jakobsen, P. R., 1987: Copenhagen Metropolitan Region Coast Erosion Management. Danish Hydraulic Institute, Hørsholm, 1–14.
Hayes, M. O., 1979: *Barrier island morphology as a function of tidal and wave regime.* In Leatherman, S. P. (ed.), Barrier Island.
Nielsen, J. and Nielsen, N., 1979: Morphology and movements of nearshore sediments in a non-tidal environment, Køge Bugt, Denmark. *Bull. Geol. Soc.* **27**, 15–45.
Nielsen, N., 1984: Shore-connected large-scale bed forms in Køge Bugt, Western Baltic, Denmark. *Geogr. Ann.* **66A**, (1–2), 121–130.
Nielsen, N. and Nielsen, J., 1985: Nearshore morpho-dynamic illustrated by volumetric fluctuations in a low energy and non-tidal environment. Proceedings European Workshop on coastal zones. Tech. Univ. of Athens, Greece, 1–15.
Recipientundersøgelser i Køger Bugt 1976–1979, 1979: VKI (Institute for Water Quality), Report to the Metropolitan Council, Copenhagen.

Niels Nielsen
Institute of Geography,
University of Copenhagen,
Denmark

16. Tourist planning along the coast of Aquitaine, France

The Atlantic coast of south-west France, particularly that of Gironde and Landes, develops in a straight, sandy littoral stretch for over 200 km from the Gironde estuary to the mouth of the Adour: this is called the Côte d'Argent for the charm of its foaming waves. The coastal zone includes various types of environments: the beach, the dunes, the 'lette', the protective forest and the exploited forest. The beach ('estran') is subject to tides, which here have a maximum amplitude of five metres. High energy breakers and westerly winds, contribute to the formation of a series of dune-belts, which rise up towards the east. In the last century, the need to stop the displacement of the sand inland, gave way to a large-scale containment operation, consisting in the imposition of an artificial profile to the dunes, with a gradual slope on the ocean side, a flat section on top, and a steeper slope on the inland side. The special vegetation cover ensures an effective fixing of the dune complex and prevents its progress inland. Beyond these dune-belts, there is the 'lette', a corridor running parallel to the coast, sheltered from the wind, and 200 m to 2000 m wide. Protected from the sea winds, the 'lette' is often an area used for settlements.

To the east, the protective forest is made up of bushy and arboreal elements, which become progressively higher, with maritime pine prevalent. It constitutes the most efficient protection for the subsequent exploited forest, which is also formed prevalently by maritime pine and, to a lesser extent, by oak.

At a few kilometres from the shoreline numerous lakes align along the back-shore. To the north, the lakes of Hourtin-Carcans (the largest in France) and of Lacanau, and to the south of the Arcachon basin (the only considerable embayment on this coast) lakes of Cazaux, Biscarrosse and Parentis, Aureilhan, Leon and Soustons, are the main ones. These are dune-dammed lakes insofar as the waterways that cross the Landes forest are prevented by the sand belts from flowing into the sea; their water level is 15–18 m above sea-level and several metres deep. In fact, a few of the waterways manage to make their way to the sea through tortuous routes ('courants').[1]

In the space of a few kilometres from the ocean, extensive beaches, large lakes, and a vast forest, constitute a successful trinomial, a slogan to publicize the touristic image of Aquitaine. This long and wide beach has, in fact, long attracted tourists, due to the wild beauty of the scenery and the unspoiled landscape.

Even before the Second World War, the Aquitaine coast was a popular summer tourist attraction in France: Biarritz and Arcachon were well-known tourist centres. After the war, tourism flourished again, and economic progress encouraged crowds of visitors. As a result, from the 1960s on, there have been conflicts regarding the utilization of space, with the creation of new kinds of reception facilities, sea-side building complexes, and new holiday resorts, often generated from small inland

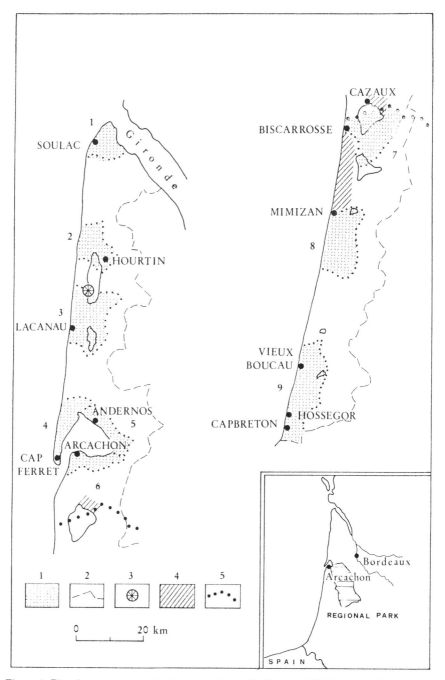

Figure 1. Planning scheme on the Aquitaine Coast. 1) The nine UPA's areas, 2) Boundaries of the MIACA operational area, 3) Base de Bombannes, 4) Military zones, 5) Department boundary.

villages. The need to coordinate various initiatives and to stop unregulated development soon became clear. Based on the findings of a work-group set up by the DATAR,[2] a decree of 20 October 1967 created the *Mission Interministerielle pour l'Aménagement de la Côte Aquitaine* (MIACA). This 'mission' had the task of defining the general program for the planning of the coast, and of determining the executive methods for implementing it. The basic principle of this plan, approved on 21 April 1972 by the *Comité Interministériel d'Aménagement du Territoire* (CIAT), was the economic and social promotion of Aquitaine. The action of the MIACA has been conducted in two stages: the first preparatory (1970–1974), in the course of which the general planning themes were defined, a land policy initiated, and various important undertakings relating to facilities, sanitation and the environment completed. The second, which continued up to 1984, was the operational one, with the implementation of projects to create the first programmed tourist resorts and the first manifestations of social tourism. The main purpose of the scheme was to control and plan the increase in the number of summer visitors to the coast. For this reason – and the originality of the plan lies in this – the Aquitaine coastal zone was subdivided into nine *Unités Principales d'Aménagement (UPA)* alternated with and separated by seven *Secteurs d'Equilibre Naturel* (SEN); thus, planning operations were concentrated in areas where tourist centres already existed and where tourist facilities and settlements could be located in a combined exploitation of sea, lakes and forest. Between these settlements, large areas of unspoiled land were interposed, where the holiday-maker could find peace and quiet, together with some facilities.

Initially, great importance was given to the construction of the trans-Aquitaine canal, which would have linked the Gironde river with the bay of Arcachon and then with the Adour, across the lakes, with the aim of modifying the tourist season, extending it from Easter and All Saints holidays. This plan was soon set aside due to the considerable technical difficulties involved.

The nine UPA's and the seven SEN's were grouped into five structures, in closer correspondence to the geographic microregions. Thus, starting from the Medoc peninsula, where tourism (Soulac and Amelie) and industrial port development (Le Verdon sur Gironde) must coexist at close quarters, we find the area of the great Gironde lakes, with tourist projects linked to the vicinity of the Bordeaux agglomeration (the Bombannes '*base de plein air*', the Lacanau-Ocean complex); then the area of the Arcachon basin, where limited increases in reception capacity, and road and environmental improvements are provided for; the area of the Cazaux and Biscarrosse lakes with the coastal resort of Biscarrosse Plage; and the most southerly area with established holiday resorts such as Port d'Albret, Hossegor and Vieux Boucau.

After the development, in the 1960s, of a concentrated tourism, favoured by the promotors of regional specialization, a 'diffuse' and more ecological tourism has subsequently emerged, which requires areas of uncontaminated natural space to allow people to live as much as possible in contact with nature. From this point of view, the Lande de Guascogne can play an important and increasing role, as about 80% of the littoral is natural space.[3] Furthermore, the straight, uniform beach is the longest in Europe and is free of urban agglomerations and of industrial and port developments. The conflict between the utilization of the area for touristic and leisure activities and the conservation of the environment has been settled in Aquitaine by a way of planning which provides for the establishment of seven 'sectors of natural equilibrium'; by the creation of nature reserves (at present there are five in Gironde and two in Landes); by the large number of 'classified sites', whose protection is guaranteed, and finally by the increasing acquisition of areas to be preserved unspoilt by the *'Conservatoire de l'Espace Littoral et des Rivages Lacustres'*.

Very close to the coast and along the bay of Arcachon, the vast Regional Nature Park of the Landes de Guascogne extends for 206 000 ha, and has the function of preserving the coherent natural complex of the Landes forest, as well as of developing inland tourism.[4] Some of its most important realisations, such as the ecology museum at Marqueze, receive more than 100 000 visitors a year. So, substantial tourist attractions are offered by this coastal tourist area, comprising the 'sun-belt' of the French Atlantic (less sunny but also much less congested than the Mediterranean). In addition, centres of *plein air et de loisirs* have been created within the context of a 'functionalized' planning of the space, i.e. a planning that conforms with the aim of giving both a response from the public authorities to the citizens' recreational needs and of performing a formative role with respect to recreational activities. These centres are animated and open to everybody. Their facilities offer the most varied possibilities for relaxation, sporting, cultural and entertainment activities in a natural, noise-free environment. In Gironde, on the western shore of the lake of Hourtin, the *Base Departementale de Sports et de Loisirs de Bombannes* has been established in 220 ha of forest. Originated as a *Base de Plein Air et de Loisirs* set up by the General Council of Gironde, and with the technical and financial assistance of the Ministry of Youth, this is a vast complex for entertainment and sport, 60 km from Bordeaux and just 4 km from the sea. Open to everyone, it is intended mainly for those who wish to take part in sporting activities. The sports facilities provided are impressive, scattered around the huge pine forest which stretches uninterrupted from the lake to the sea.[5] The success of this project has encouraged the Council of Gironde to plan another, similar centre, including facilities for the very young, at Hourtin-Port, on the north-east shore of the same lake.[6] The littoral and lacustrine areas of southern Gironde benefit from the vicinity of Bordeaux; this same factor also constitutes a valid reason for the location of industries in the area to the west of Bordeaux, thus creating an interference between the urban and the coastal tourist systems.

The demand for leisure areas is growing, due to the increase in free-time, and to increasing specialization of work. Due also to the influence of the mass-media, 'naturism' has spread to all beaches. Multiple use and planning problems have become associated with these motivations, so that MIACA has directed the growth of new leisure zones towards a kind of tourist development that makes the best of nature.[7] On the Aquitaine coast touristic/recreational exploitation has been planned along lines of direction perpendicular to the beach, starting from the sea and reaching the lakes (sailing and surfing are very widespread there) and into the forest (ideal for cycling, walking and running, canooing on the 'courants').

Anticipating a sizeable increase in tourism along the whole coast, the *Mission* worked out the plan summarized above with the expectation of a notable expansion in the hotel sector and in holiday homes. In reality, figures for the last few years show that this anticipated increase has not taken place. Growth has been slowed down by the economic and energy crises. The forecasts of distribution among various sectors have also been changed: increase in hotels has been very limited, whereas there has been a notable increase in camping sites in particular, and also in holiday homes, despite the fact that these were discouraged in order to avoid a large number of constructions uninhabited for most of the year.

Total days spent by summer holiday-makers in hotels and camping sites on the Gironde and Landes coasts increased from five and a half million in 1974 to more than eleven million in 1986. Since tourism/recreation is one of the developing activities, offering a substantial revenue, it is one of the fields of action favoured by the Regional Office, which has transferred to the MIACA, transforming it into a specific organization for tourist planning in Aquitaine. Since 1985, the local government of Aquitaine has committed itself to the preparation of the Integrated Mediterranean Programs in order to prevent difficulties caused by the inclusion of Spain and Portugal in the EEC. Tourist programs account for 12% of the budget and are directed towards various types of action: the creation of new reception facilities throughout the region; the characterization of the region by highly prestigious tourist attractions (re-exploiting old resorts such as Arcachon and Biarritz, and expanding new coastal resorts as Plan Golf); and the exploitation of Aquitaine as a tourist attraction in the European context.[8]

In effect, the wide stretch of Aquitaine coast constitutes a single system including various interacting and synergy-developing elements: Bordeaux-lakes-forest-sea; i.e. a vast metropolitan area – the main source of the tourist demand – a lacustrine reception zone equipped with entertainment and sports centres, an extensive area of forest particularly suitable for relaxation, and a sandy coast – a paradise for holiday-makers in search of sea and sun. All these areas are interconnected and constitute a system open to external relations.

A large mass of summer tourists along a coast undoubtedly creates problems of conservation of the environment.The increase in holiday homes and camping sites has created a growing demand for land and accommodation on the sea-shore, causing further difficulties for the land policy and for the occupation of favoured sites. This massive summer population creates difficult problems for structures related to hygienic conditions, waste disposal, water supply and traffic. Only the water supply problem has been completely resolved here, largely because it responds to the need for defence against fires, so dangerous in the Landes forest. However, the greatest danger to the environment comes from the trampling of the sand dunes by recreationists. We noted earlier that the coastal dunes of the Landes de Guascogne have an artificial morphology, modelled on the ideal shape to resist deflation. The permanence of this shape is tied to the presence of the psammophilic vegetation (thorny esparto or '*gourbet*'), which holds the sand together, or to the covering of vegetable residue. The trampling of the dune surface destroys the '*gourbet*', so that the sand, no longer held together, is blown away by the wind. Deflation channels form which gradually produce furrows in the dunes,which can in turn lead to the formation of sand heaps in the *lette*, and then affect the protective forest. This explains the policy for safeguarding the coastal dunes by means of the *Plan Plages* of Gironde and Landes, which were worked out by the MIACA and by the *Office National des Forêts* to improve tourist reception but also to protect the environment. For the beaches in particular, the intention in many areas is to establish controlled and limited access and plantations of shrubs and esparto, as well as ways of educating and informing the users. Other problems require appropriate action. For example, the retreat of the beaches on the Landes de Guascogne coast explains the large number of initiatives and research projects aimed at reducing this phenomenon. Similarly, a great deal of attention is being directed to the safety of the bathers.

To resolve these problems, an attempt has been made in the Landes de Guascogne to stop the process of concentration that the nature of the coast induces (one-dimensional space) by directing a large amount of touristic/recreational activity inland, thus limiting the development alongshore and giving it a second dimension, spreading the tensions 'orthoganally', and rebalancing (or at least beginning to rebalance) the coast/inland space system. All this to ensure that the areas dedicated to leisure activities remain qualitatively acceptable.

One merit of the administrative bodies that have set out the plans for the Aquitaine coast is that of working with a certain flexibility, which manages to maintain a link between the changing requirements of the tourists and the need to protect the environment. This policy is most obviously demonstrated in the Biscarrosse area (unit 7), where initially the MIACA planned a tourist development centred on a lakeside village, built on the trans Aquitaine canal and with a large number of facilities. Once the canal project had been reconsidered and the idea of enlarging the lakeside village consequently abandoned, the municipality of Biscarrosse focused on the gradual development of Biscarrosse Plage. This seaside resort has, in fact, registered a slow but constant increase in camping sites and family houses; but it has met the requirements of a tourism which is more and more respectful of the environment, without causing damage to the seashore. The success of this change of plan is shown by the growing interest in this resort, which is being pleasantly updated with facilities and driving its expansion towards the area behind the *lette*. The central part of the dunes, where Biscarrosse Plage faces the sea, has been admirably arranged, using funds made available by the Regional Council as part of the policy of re-exploitation of the old resorts. Development of reception facilities at the Biscarrosse Bourg lake has not, however, been ignored and has focused on camping sites and family accomodations.

In conclusion, in Aquitaine, the attraction of the beaches, a considerable number of sunny days and a wave regime suitable for surfing; abundant possibilities for sailing and associated sports on the numerous inland lakes; the considerable leisure potential of the great Landes forest, constitute a recreational setting of considerable interest which has undergone a constant evolution in synthony with such social processes as urban expansion and cultural evolution. The planning initiatives which have been carried out in this region have had the merit of renewing its attraction, concentrating operations and tourist facilities in restricted areas, leaving large areas of countryside untouched, often adapting the *'aménagement'* to the changing requirements both of the holiday-makers and of environmental protection with appreciable flexibility.

Notes

1. M. Cassou-Mounat, 1977, *La vie humaine sur le littoral des Landes de Gascogne*, Paris, Champion.
2. DATAR = Delegation à l'aménagement du territoire et à l'action régionale.
3. 'Natural spaces' means uninterrupted areas, 2 km long and 2 km wide, so that on the Aquitaine coast there are at least 152 km of this kind. B. Wagon, *Le littoral aquitain: Paysage et Architecture*, Libourne, 1986, Arts graphiques d'Aquitaine, p. 181.
4. P. Ghelardoni, 1984, *Alcune note sul Parco regionale delle Lande di Guascogna, Boll. Soc. Geogr. Ital.* 7–9.
5. 26 tennis courts, an Olympic swimming-pool and one for children, a sailing school and a surfing school, 2 gymnasiums, a mini-golf course, an archery range, a canooing centre, a network of cycle tracks, an orienteering course, a Basque-ball court, 3 supervised beaches; in addition, a culture centre and a museum of folk crafts and traditions. As for reception facilities, there are 3 camp-sites, a picnic area, low-cost accommodation, car parks etc.
6. In this case, however, the operation has been entrusted to a mixed local/regional syndicate, with a large private-sector component.
7. J.-P. Augustin, 1986, *Pratiques de la mer et territoires urbaines: de nouveaux espaces de loisirs sportifs pour l'agglomeration de Bordeaux, R.G.P.S.O.* **4**, 589–609, cfr., p. 607.
8. J. Dumas, 1986, *Compensation ou animation: le tourisme dans les interventions de la région Aquitaine, R.G.P.S.O.* **4**, 573–587, cfr., p. 578.

Paolo Ghelardoni
Institute of Economic Geography,
University of Pisa,
Italy

17. Sydney's southern surfing beaches: characteristics and hazards

Introduction

The Sydney metropolitan coast extends for 58 km in north-south direction, centered on latitude 34° and fronts the Tasman Sea (Figure 1). The coast consists of prominent sandstone and shale headlands, separated by 25 bay-head and mid-bay beach-barrier systems (Short and Wright, 1981). All of these beaches are utilized by the public throughout the year, but particularly during the warmer summer months (October–April) when hundreds of thousands may flock to the beaches on a hot day. Because of their long and continued popularity, the moderate to occasionally high breakers and common rips on most beaches (McKenzie, 1957; Short, 1985) public safety has long been a major component of beach management in Sydney. The world's first surf lifesaving association was formed in Sydney in 1907. Sydney now has the world's greatest density of volunteer surf clubs, with 33 clubs on 19 of the beaches. In addition to the voluntary surf lifesavers, council lifeguards also patrol the beaches. All beaches are patrolled daily between October and April and some year round. On an average year over 2 000 000 people bath on the beaches of whom between 4000–8000 will be rescued from the surf (Hogan, 1987), while 5–10 fatalities will occur, primarily due to unexperienced people being caught in rips. Despite the great popularity of surfing in Sydney and around the temperate and tropical world no studies have yet been undertaken to access the relationship of beach type and potential hazards and of public awareness of hazards. This paper reports on a preliminary investigation of six beaches patrolled by eleven surf clubs on Sydney's southern beaches (Bondi to Cronulla, Figure 1). The study is part of a New South Wales (N.S.W.) state wide 'Beach Grading and Hazard Assessment' being undertaken by the authors in conjunction with the Surf Life Saving Association, N.S.W. State Centre, involving all 124 patrolled and 307 unpatrolled beaches along the State's 1170 km of sandy coast.

The beaches

The study involved six beaches which are patrolled by eleven surf clubs. The general beach and surf zone characteristics under modal wave conditions ($H_o = 1.5$ m, $T = 10$ sec, $D = SE$, Short and Wright, 1981) is illustrated in Figure 2, together with the location of the surf clubs and major permanent features. The beaches range from those most exposed to the dominant east and south east waves and which experience the highest breaker waves (Bondi, Tamarama, Bronte, Maroubra and Wanda) to those afforded increasing protection by headlands and reefs and which experience increasingly lower waves (North Bondi, South Maroubra, South Cronulla, Coogee and Greenhills, see Figure 2 and Table 1).

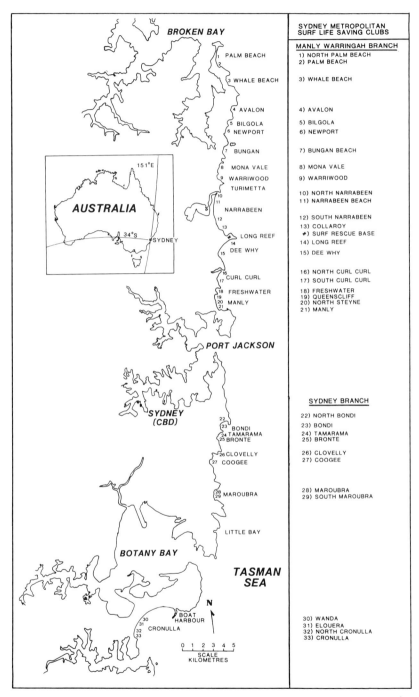

Figure 1. The Sydney metropolitan coast extends 58 km from Broken Bay to Port Hacking. It includes 27 beaches of which 19 are patrolled by members of 33 surf clubs and council lifeguards.

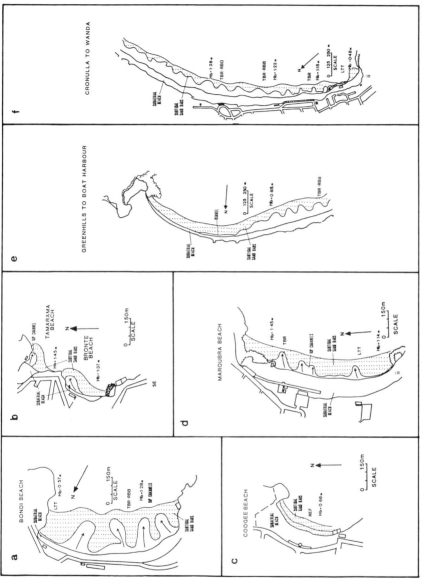

Figure 2. Sketch maps of the six southern Sydney beaches, illustrating the typical summer beach and surf zone morphology, including modal breaker height (Hb) and beach type (rhythmic bar and beach – RBB; transverse bar and rip – TBR; low tide terraces – LTT; reflective – R) (Modified from Hogan, 1987).

Field data

The study is based on three data bases. First field data obtained from each beach between January 10 and April 25, 1987. Second, analysis of surf club records of daily beach population and rescues between 1949 and 1986; and third, analysis of 13 sets of aerial photographs of the beaches taken between 1951 and 1986 to access temporal changes in beach and surf zone morphology. The details of each are contained in Hogan (1987). Field data consisted of sketching of beach and surf zone morphology on 42 occasions, including breaker conditions, location of rips, bars, etc; of representative beach and surf zone cross-sections across each beach; and of 496 interviews of beach users as to their awareness of hazards and lifeguards.

Results

The results can be accessed in three ways – the physical nature of each beach and its spatial and temporal variability in response to changing wave conditions; the hazards associated with each beach both in potential and real terms; and the public awareness of these hazards. Only the first two will be addressed in this brief report.

Table 1. Physical, bathing and rescue characteristics of the twelve patrolled sections of the six southern Sydney beaches.

	Hbs m	B/S	Ys m	Psw x 1000 persons	R persons	RPT
North Bondi	0.8	LTT/TBR	240	300	141	0.47
Bondi	1.3	TBR/RBB	180	390	213	0.54
Tamarama	1.5	RBB/RBB*	70*	24	145	5.56
Bronte	1.4	TBR/RBB*	140*	60	116	1.93
Coogee	0.7	R	0	25	41	0.43
Maroubra	1.5	TBR/RBB	200	80	268	3.41
South Maroubra	1.2	LTT/TBR	210	28	88	3.17
Greenhills	0.9	TBR/LTT	230	-	-	-
Wanda	1.4	TBR/RBB	250	42	71	1.71
Elouera	1.3	TBR/RBB	270	35	124	3.53
North Cronulla	1.2	TBR	180	35	174	5.02
Cronulla	0.5	LTT	130	57	118	2.08

Hbs – modal summer breaker wave height (Hogan, 1987); T = 10 sec: B/S – summer beach state (see Figure 3 and Wright and Short, 1984); Ys – rip spacing; Psw – annual summer weekend beach populations (1985–87); R – annual number of rescues (1949–1986); RPT – rescues per thousand = R/Psw;
* indicates headland influence on B/S and Ys.

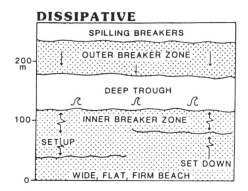

DISSIPATIVE

Characteristics
* dissipative – waves dissipate energy over a wide surf zone
* fine sand, 2–3 m breakers, straight bars, trough and beach
* only occur in N.S.W. during high waves on fine sand beaches, during cyclones in south Queensland, more common on south coast of Australia

* Outer bar: 1–2 m deep, spilling breakers
* Trough: 2–3 m deep, possible longshore flow
* Inner bar-terrace: (spilling-plunging breakers possible transient rips
* Swash zone: set up and swash bores and set down and return flow (every 1–2 minutes), can trap unwary (Note: Swash zone is area where waves run and down beach).

Hazards
* high waves and wide surf zone restrict most bathers to swash zone
* safest bathing – in swash zone, with care of set up and set down
* hazard rating – 6/10 (most people stay close to shore)

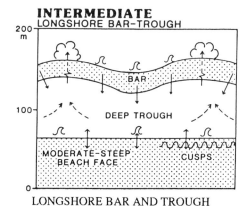

LONGSHORE BAR AND TROUGH

Characteristics
* consists of shore parallel bar and trough

Figure 3.

* fine-medium sand, 1.5–2 m breakers, weak to moderate rip currents
* common in south Queensland, northern N.S.W., south coast of Australia

* Bar: plunging waves (low tide), weak to moderate rips (200–300 m apart)
* Trough: 2–3 m deep, weak to moderate longshore and rip currents
* Swash zone: moderate to steep beach face, shorebreak especially at high tide
 (Note: intermediate refers to intermediate between dissipative and reflective beach types)

Hazards
* deep trough and distance to outer bar restrict most bathers to swash zone and inner trough
* safest bathing – swash zone and in trough away from rip
* hazard rating – 7/10 (most people stay close to shore)

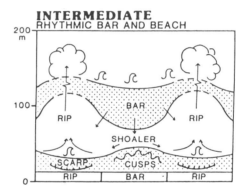

RHYTHMIC BAR AND BEACH

Characteristics
* consists of rhythmic (undulating) bar and trough, with straight beach
* fine-medium sand, 1.5 m breakers, distinct rip troughs separated by detached bars
* common south Queensland, N.S.W.(after high waves and on more exposed beaches), south coast of Australia
* Bar: plunging waves on bar at low tide, moderate to strong rips every 150–300 m
* Trough: alternates from shallow with onshore currents (behind bar) to deeper with pulsating rip feeder and rip currents
* Swash zone: lower-waves and shallow water behind bars;
 shorebreak, deep water and feeder currents behind rips.

Hazards
* pronounced changes in depth and currents between bars and rips
* safest bathing: on or behind bars during lower waves, hazardous during high waves and at high tide
* hazard rating – 8/10 (people wade or swim to shoaler bars, bathers caught or washed into rips).

Figure 3 (*continued*)

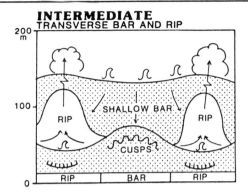

TRANSVERSE BAR AND RIP

Characteristics
* consists of attached bars, rip troughs and undulating beach
* fine-medium sand, 1–1.5 m breakers, distinct rip troughs, separated by attached bars every 150–300 m
* most common in south Queensland, N.S.W., inner bars on south coast of Australia

* Bar: welded hence transverse to shore, shallow with plunging waves, alternate with deeper (1–2 m) rip troughs and currents
* Trough: occupied by rip feeder currents and rips
* Swash zone: shallow with low waves in lee of bars; steeper beach and low shorebreak in lee of rips

Hazards
* pronounced change in depth and currents between bars and rips
* safest bathing – on bars
* hazard rating – 8/10 (people bathe on shallow bars adjacent to rips, however bathers can be swashed off bar into rip, inexperienced bathers enter rips).

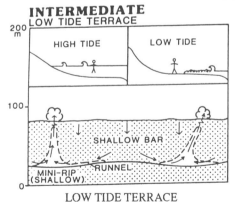

LOW TIDE TERRACE

Characteristics
* has shallow bar or terrace often exposed at low tide

Figure 3 (*continued*)

* fine-medium sand, 0.5–1 m breaker
* occur toward lower wave section of ocean beaches
* relatively uniform alongshore, apart from small rips

* Bar: low tide – very shallow, plunging breakers
* Trough: shallow rip trough and weak currents may cross bar
* swash zone: moderate to steep beach, dry at low tide, shorebreak at high tide

Hazards
* safest bathing – safe at low tide, deeper and weak rips at high tide
* hazard rating – 4/10

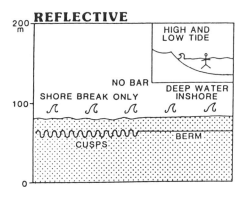

REFLECTIVE

Characteristics
* reflective – waves tend to reflect back off beach
* medium to coarse sand, 0–1 m breakers
* only occur on very low wave beaches and on harbour beaches
* uniform alongshore apart from cusps (20–30 m spacing if present)

* no bar
* no trough, though 1–2 m deep water close inshore
* Swash zone: steep beach face with low shorebreak

Figure 3 (*continued*)

> Hazards
> * safest bathing – safe apart from deep water close inshore and from shorebreak during high waves. Steep beach and abrupt drop off to deeper water can make access difficult for elderly and children
> * hazard rating – 3/10
>
> Please note:
> The above represent average waves and conditions (as indicated) on these beach types in micro tidal (<2 m tide range) regions of southern Australia (south Queensland, N.S.W., Victoria, Tasmania, South Australia and southern Western Australia).
>
> The hazardousness of each type is increased by:
> Headlands – rips usually occur adjacent to headlands, reefs and rocky outcrops
> Oblique waves – stronger longshore currents, skewed and migratory rips
> High tide – deeper water and in some cases stronger rips
> Rising seas – eroding bars, stronger currents, strong shifting rips, greater set-up and set down
> High tide and rising seas – more difficult to distinguish bars and troughs
> Strong onshore and alongshore winds – reinforced downwind currents
> Megaripples – large migratory sand ripples common in rip troughs can produce unstable footing
> Changing wave conditions – (rising, falling, change in direction or length) – produce a predictable change in beach topography and type, the reason why beaches are always changing.
>
> Also note that sandbars can not and do not collapse; and no waves are 'freak' some are just a little bigger than others.

Figure 3. Beach types and associated characteristics and hazards to bathers. Adapted from Wright and Short, 1984. Produced by the Coastal Studies Unit, University of Sydney. Concept and design: Andrew Short and Christopher Hogan, 1988.

Beach characteristics

Figure 2 and Table 1 provide an overview of the modal characteristics of each beach. North Bondi is partially protected by the northern headland and has a modal breaker height of 0.8 m and a LTT to TBR beach (Figure 2a). More exposed Bondi with a $Hb = 1.3$ m has a TBR-RBB beach type (Figure 2a). The small Tamarama and Bronte beaches are both exposed to the dominant SE waves with $Hb = 1.5$ and 1.4 m respectively and TBR-RBB beach types (Figure 2b). Their surf zones are however severely modified in terms of rip location by the dominance of headland controlled rips. Coogee Beach is protected by offshore reefs and an island and has the lowest Hb of 0.7 m and is modally reflective (Figure 2c). Maroubra Beach is more exposed toward the north with a Hb of 1.5 m and TBR-RBB beach type, while slightly more protected and lower energy South Maroubra has a $Hb = 1$ m and TBR beach type (Figure 2d). The unpatrolled Greenhills beach experiences increasing northward protection from an offshore reef with waves decreasing from $Hb = 0.9$ to 0.5 m and beach type from TBR to LTT (Figure 2e). Wanda, Elouera, North Cronulla and Cronulla illustrate a similar longshore variation in breaker height (1.4 to 0.5 m) and resulting change in modal beach state from RBB to TBR to LTT (Figure 2f).

Beach hazards

The potential and actual hazards associated with each beach can be gauged by a comparison with the actual beach type (Figure 2) and the hazards associated with each shown in Figure 3, and from the rescue records in Table 1. In terms of potential hazards those beaches with the highest waves and largest most numerous and most frequently occurring rips are the most hazards. As Figure 3 indicates these are associated with the RBB and TBR types which contain the ingredients of shallow bars adjacent to deeper troughs containing rips, and swash zones adjacent to deeper troughs and rip feeder channels, in addition to the moderate breakers (1–2 m) and stronger currents and rips. On the south Sydney coast beaches dominated by the RBB-TBR beach type, include Bondi, Tanamara, Bronte, Maroubra, Eloura and Wanda.

Higher energy beaches (D and LBT) whilst potentially more hazardous (Figure 3) only occasionally occur on the Sydney coast, and when they do occur are self regulatory as either the beaches are closed to the public by the lifeguards and/or the public clearly sees the high hazardous waves and surf and stays out or in shallow water.

The lower energy beaches (LTT and R) are the safest surfing beaches owing to their uniform longshore nature, low waves and general absence of rips and strong currents (Figure 3). On the south Sydney coast North Bondi, Coogee, Greenhills and South Cronulla are regarded as the safest beaches as a result of their beach type (Table 1).

In order to test the relationship between potential and actual hazards surf club beach population records (1985–87) and rescue records (1949–86) were assessed. Unfortunately long term reliable beach population records are not available. Given this limitation in the data we still feel it is useful to compare these data sets to obtain an approximate measure of actual hazards. These are summarized in Table 1. The most popular beach is Bondi with an average annual population of 690 000, with Coogee and the very small Tamarama being the least popular. However in terms of absolute rescues Maroubra is the most hazardous with an average of 270 rescues per year, and Coogee the least with 41 rescues per year.

In order to standardise these results for beaches with significantly different populations the number of rescues per thousand people (RPT) was calculated for each patrolled beach (Table 1). These results indicate that Tamarama is the most hazardous with 5.6 RPT followed by North Cronulla, (5.02), Eloura (3.53), Maroubra (3.41) and South Maroubra (3.17). All these beaches with more than 3.0 RPT are modally TBR to RBB the most hazardous beach types (Figure 3). The remaining beaches with fewer rescue include Cronulla (2.08), Bronte (1.93), Wanda (1.71), Bondi (0.54), North Bondi (0.47) and Coogee (0.43). These beaches include the rip dominated TBR as well as the two LTT and one reflective beach. This indicates that lower energy, less hazardous beaches tend to experience fewer rescues per thousand. Bronte is a potentially hazardous beach which has a safer than expected record probably owing to the three swimming pools adjacent to the beach (Tamarama has none) and ease of patrolling and monitoring safety on a small beach (Figure 2b).

Naturally many other factors influence the number of rescues on a particular beach including the age, sex, background and surfing experience of the bathers, the proportion of sunbathers to actual bathers and so on, however the above figures do indicate that there is a positive correlation between increasing wave height and beach type with the number of rescues per thousand on the southern Sydney beaches.

Discussions and conclusions

Sunbathing and bathing on the world's tropical and temperate ocean beaches is becoming an increasingly popular pastime and recreational pursuit. Increasing global affluence, leisure time and tourism is likely to accelerate this trend. There is therefore an urgent need to ensure that public safety is maintained at recreational and tourist beaches. Whilst techniques for rescuing bathers are well advanced in many countries, there has been little application of our growing scientific knowledge of beach and surf zone dynamics to the problem of recognition and management of beach hazards.

This study of six popular surfing beaches patrolled by eleven surf clubs on Sydney's southern coastline reveals that there is a positive correlation between the number of rescues per thousand beach users and the wave height and beach types. Most rescues occur on moderate wave energy beaches dominated by rips, and least or lower energy more uniform, ripless low tide terrace and reflective beaches. This is also confirmed by an event analysis by Hogan (1987) which showed highest number of rescues occurred on days when waves exceeded 1.6 m on TBR-RBB beaches and when beaches were crowded (warm to hot, sunny weekends and holidays). Additionally each of the six beach states of Wright and Short (1984) has a characteristic set of hazards (Figure 3) that can be used in a variety of ways, including public education, prediction of beach change and hazards (Wright et al., 1984; Short, 1987) and management of public beaches.

Acknowledgements

This project was supported in part by Randwick Municipal Council. It was conducted in cooperation with the Surf Life Saving Association, N.S.W. State Centre with assistance from Andrew May and the eleven surf clubs listed.

References

Hogan, C. L., 1987, Public hazard and awareness study of Sydney's southern beaches. B.Sc. Hons. thesis (unpubl.) Dept. Geography, University of Sydney, 289.
McKenzie, P., 1958, Rip-current systems. *J. Geol.* **66**, 103–113.
Short, A. D., 1985, Rip-current type spacing and persistence, Narrabeen Beach, Australia. *Mar. Geol.* **65**, 47–71.
Short, A. D., 1987, A note on beach type and change, with examples from S.E. Australia. *J. Coastal Res.* **3**, 387–395.
Short, A. D. and Wright, L. D., 1981, Beach systems of the Sydney region. *Aust. Geogr.* **15**, 8–16.

Wright, L. D. and Short, A. D., 1983, Morphodynamics of beaches and surf zones in Australia. In Komar, P. D., (ed.), *Handbook of Coastal Processes and Erosion*, CRC Press, Boca Raton, 35–64.

Wright, L. D. and Short, L. D., 1984, Morphodynamic variability of beaches and surf zones: A synthesis. *Mar. Geol.* **56**, 93–118.

Wright, L. D., Short, A. D., and Green, M. O., 1984, Short term changes in the morphodynamics states of beaches and surf zones: an empirical predictive model. *Mar. Geol.* **62**, 339–364.

A. D. Short and C. L. Hogan
Coastal Studies Unit,
Department of Geography,
University of Sydney,
Australia

18. Twenty five years of development along the Isreali Mediterranean coast: goals and achievements

Introduction

The Israel Mediterranean coasts form the southeastern part of the Levantine Basin. They are smoothly curved, forming a large arc that originates in northern Sinai and continues toward the Mt. Carmel promontory (Figure 1). These beaches belong to the edge of the Nile littoral cell that commences at the Nile River outlets at Rashid (Rosetta) and Dumyat (Damietta). The main net sand-transport in this cell is from the Nile Delta to north Sinai and the Israeli beaches, terminating at 'Akko, northern Israel. Sand quantities gradually decrease from the sources towards the edge of the cell. Quartz is the main component of the sand, with some calcium-carbonate (broken-shells of local origin), and a wide range of heavy minerals. On the other hand, the 'Akko – Rosh Haniqra beaches belong to an independent littoral cell, and differ considerably in their mineralogy. Calcium carbonate grains, mostly of local sources, are the main components of this isolated cell, while quartz of Nilotic origin is represented in minute quantities.

With the exception of Haifa Bay, which morphologically results from the promontory formed by Mt. Carmel, the Israeli shoreline has no pronounced bays that provide shelter and quiet water. As a result, the 240 km long beach is exposed during the summer to westerly and southwesterly waves originating from the long fetch of the Mediterranean Sea. The sea is relatively rough during the high bathing-season, and loss of life on these beaches is high. Forty four drownings were recorded during the 1986 bathing season along the Israel Mediterranean beaches. Measures were therefore taken in order to create shelter and quiet water for the bathers. Groins, detached breakwaters, closed structures, and other constructions were built. The results of these constructions in the coastal region are discussed.

These structures resulted in the development of larger beaches, with some areas of quiet water on the one hand, but with many serious problems for the beaches and bathers on the other. *Twenty five years' construction of offshore structures improved to only a small extent the relatively bad bathing conditions of the Israel Mediterranean coast, although they have, to a large extent expanded the coastal area in the heavily populated regions.*

Offshore structures

About 30 structures were built for recreational purposes since 1965, (all composed of limestone and dolomite blocks), mainly next to large urban centers (Figure 1). These structures include detached breakwaters, groins, sea-walls, marinas, and small ports (Figure 2). The early-built detached breakwaters were very long in

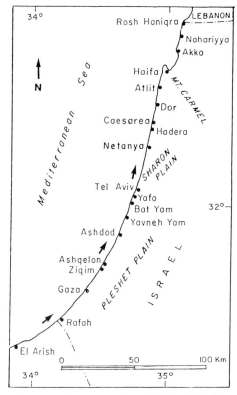

Figure 1. Location map of structures along the Israel Mediterranean shores.

Figure 2a. Ziqim northern groin. Note the sand accumulation on the left (south) side of the structure (6.8.76).

Figure 2b. Hilton detached breakwater, Tel Aviv. A nice bay was developed between this structure and its twin to the north (right). Tel Aviv marina is at the far side of the photo (27.10.74).

Figure 2c. Tel Baruch offshore structure, north of Tel Aviv (19.4.71).

relation to their distances from the original shoreline. They thus did not fit with the empirical rule, dealing with the structure's distance to length relations (Inman and Frautchy, 1966; Nir, 1976, 1982; Rosen and Vajda, 1982). As a result, *huge amounts* of sand were accumulated between the beach and the breakwaters, connecting in most cases the former to the latter by a trapezoidal or triangular-shaped sand-body called 'tombolo' (Figure 3). The effective water area was very small and there has practically been no improvement in bathing conditions. These tombolos develop to maturing over an average period of 4–5 years (Nir, 1982a, 1982d). In order to avoid large sand accumulations, structures built in later periods

Figure 2d. Tel Aviv marina, at full capacity, after about 10 years of operation (30.8.83).

Figure 3. Well-developed tombolo at the 'shadow' of Carmel Beach detached breakwater (photo by Survey of Israel, Ministry of Labour).

were located farther offshore, or were shorter in length, resulting in smaller accumulation of sand and smaller areas of quiet water.

There is no doubt that these offshore structures, constructed at any orientation to the original shoreline, intervene with longshore sand transport which mainly has a net northerly trend (Emery and Neev, 1960; Manoujian and Migniot, 1975; Inman et al., 1976; Golik and Goldsmith, 1979; Golik and Goldsmith, 1980; Nir, 1980,

Figure 4. Severe erosion of beach and small-cliff north and next to the Ziqim harbour at the end of summer (when the beaches are the widest!). These beaches, in 'prior to construction era' were very wide all year long, suffer since the construction of non-stop erosion.

1982c, 1985; Rosen and Vajda, 1982; Rohrlich and Goldsmith, 1984; Inman and Jenkins, 1984; Carmel et al., 1985). The usual phenomenon occurring in the vicinity of such offshore structures is distinctive accumulation of sand on the 'upcurrent' side of the sand transport direction, and a very pronounced erosion on the 'downcurrent' side of the structure. This erosion results from both a sand deficit and wave refraction on the northern edge of the breakwater. Beaches become narrower, the natural beach and backshore protection is minimized, and the whole system enters a triggered cycle of accelerated erosion. Specific conditions in central Israel, where the beaches were originally narrow, were even worse than those described, due to the enormous quantities of sand that were quarried along the beaches up to 1964.

Bathing conditions

The original aim of the offshore detached breakwaters have achieved some of the declared goals. Beach areas were expanded, quiet water exists to a large extent during the bathing season (in some sheltered areas in the 'shadow' of the structures), and the sizes of sand-beaches in relation to the length of the original shoreline has been extended. On the other hand some very severe problems have arisen: the stone and concrete structure is dangerous to swimmers both in the sheltered region and in the open sea; also fishermen using the stony platform are at risk during stormy days. Swimmers drownings are due both to the turbulent currents, holes in the ground, and to the 'psychological' feeling of a safe shelter in the 'shadowed' regions. Figures for the years 1978–1986 for Ashqelon beach show that casualties rose from 1–3 drownings in pre-structures era to 4 after construction. The totals for the entire Israeli coasts show practically no improvement in beaches with sheltered areas, as compared with beaches without any offshore structure. There is therefore no doubt that construction of offshore structures should be planned only in regions where conditions are very rough and no other methods can be implemented. In other regions, structures should be either improved or not constructed at all.

Marinas

At present, Israel has one 'official' marina, built in the early 1970s, near Tel Aviv, next to a system of detached breakwaters. The marine extends some 250 m offshore, with an entrance depth of over 3 m. It suffers from permanent silting which requires seasonal dredging. With overcrowding, demands for more marinas arose, and old harbours such as Jaffa (Yafo) and 'Akko were implemented as marinas. Although anchoring areas have increased to some extent, the demand for more and better marinas has led government and municipal planners toward the construction of a series of new marinas. At present, five plans are at an advanced stage: 'Akko, Herzliyya, Jaffa, Ashdod and Ashqelon. Other municipalities (Caesarea, Haifa, Bat Yam, Netanya and Nahariyya) are also included in the long range plans.

As noted above, most of the Israeli Mediterranean beaches suffer from a sand deficit (i.e., erosion) and during the winter are very narrow. The effect of medium-sized marinas which extend some 350–450 m offshore, may result in very severe erosion of both backshore and the cliffs of the 'downcurrent' side of the marina. Thus, contemporaneous construction of a series of marinas would be fatal to Israeli beaches. The accelerated severe shore erosion at beaches near Ashqelon (Figure 4), results directly from the small harbour. A series of recently constructed 3 offshore breakwaters adds enormous effect to the overall beach problems and erosion in the region. It is therefore suggested that planning and construction of new marinas should slow: old harbours with additional improvements and enlargements should be implemented first; and only later, new marinas can be erected at a small and controlled speed. The following table summarizes the prevailing conditions resulting from the offshore structures along the Israel Mediterranean coasts.

Acknowledgements

The authors wish to thank the municipal life-savers for their information on the structures and bathers along the coasts of Israel. Thanks are given to Dr. Ron Bogosch of the Geological Survey of Israel who has improved the English of this paper.

Table 1. Changes of various nature that have occurred in and around the offshore structures.

Feature	Improvement	Problems
Length of beaches	Large improvement	No problems
Near-by beaches	Accumulation in the 'upstream' side	Accelerated erosion in the 'downstream' side
Far-away beaches	General deficit of sand, removed from the system by sand bodies, (tombolos, etc.)	Affect narrow beaches and accelerated shore and backshore (mainly cliffs) erosion
Size of beaches	Expanded of sand area in relation to original shoreline	Some unused areas at the central parts of the tombolo
Bathing conditions	Small extent	Many problems, mainly along and next to edges of breakwater: turbulent currents, etc., causing casualties
Stagnation	None, under natural conditions	Beach and water stagnation, during summer, as a result of large accumulation of algae
Breakwater	Good deep water 'platform' for fishermen and sun-bathers during quiet seasons	Dangerous during storms, unfit stones and concrete floor cause severe injuries

References

Carmel, Z., Inman, D. L., and Golik, A., 1985. Directional wave measurement at Haifa, Israel, and sediment transport along the Nile littoral cell. *Coastal Eng.* 9, 21–36.
Emery, K. O. and Neev, D., 1960. Mediterranean Beaches of Israel. *Bull. Geol. Survey of Israel* 26, 24.
Fried, I., 1975. Foreshore and beach development of Tel Aviv and Natanya. Symposium on: Foreshore and beach development from the coastal engineering aspect. Assoc. of Eng. and Archit. in Israel, Tel Aviv; p. 9 (in Hebrew).
Fried, I., 1976. Coastal protection by means of offshore breakwaters. Proc. of the 15th. Coastal Eng. Conf., Honolulu, Hawaii; p. 10.
Goldsmith, V. and Golik, A., 1978. The Israeli wave climate and longshore sediment transport model. Isr. Ocean. Limnol. Res., Geol. Dept. Report No. 78/1; p. 56.
Goldsmith, V. and Golik, A., 1980. Sediment transport model for the southeastern Mediterranean coast. *Marine Geology* 37, 147–175.
Inman, D. L., 1978. The impact of coastal structures on shorelines. Coastal Zone, *Amer. Soc. Civil Eng.* 1, 2265–2272.
Inman, D. L., Aubery, D. G., and Pawka, S. S., 1976. Application of nearshore processes to the Nile Delta. In: Proc. seminar on Nile Delta Sedimentology, Alexandria; UNDP/UNESCO Project for Coastal Studies; pp. 205–255.
Inman, D. L. and Jenkins, S. A., 1984. The Nile Littoral Cell and Man's Impact on the Coastal Zone of the Southeastern Mediterranean, Scripps Inst. Oceanog., Reference Series 84-31; p. 43.
Manoujian, S., and Migniot, C., 1975. Sedimentological study in three dimensional Model. Lab. Centr. Hydrol. France, Report, p. 57.

Nir, Y., 1976. Detached breakwaters, groins and other marine constructions and their influence on the Israel Mediterranean Beaches. Geol. Survey of Israel Report MG/2/76; p. 33. (in Hebrew).

Nir, Y., 1980. Recent sediments of the Haifa Bay. Geol. Survey of Israel. Report No. MG/11/80; p. 8.

Nir, Y., 1982a. Offshore artificial structures and their influence on the Israel and Sinai Mediterranean beaches. Geol. Survey of Israel, Report MGG/4/82; p. 24.

Nir, Y., 1982b. Asia, Middle East, Coastal Morphology: Israel and Sinai. In *Encyclop. of Beaches and Coastal Environments*, M.L. Schwartz (ed.), Hutchinson, Ross Pub. Co., Stroudsburg, Pennsilvania; pp. 86–98.

Nir, Y., 1982c. Israel. Chap. 70 In *The World's Coastline*. Bird, E. C. F., and Schwartz, M. L. (eds.); Van Nostrand Reinhold Co., New York; pp. 505–511.

Nir, Y., 1982d. Offshore artificial structures and their influence on the Israel and Sinai Mediterranean Beaches. 18th. Coastal Eng. Conf., ASCE/Cape Town, S. Africa. pp. 1837–1856.

Rohrlich, V. and Goldsmith, V., 1984. Sediment transport along the Southeastern Mediterranean: A Geological Prospective. *Geo-Marine Letters* 4, 99–103.

Rosen, D. S. and Vajda, M., 1982. Sedimentological influence of detached breakwaters. 18th. Coastal Eng. Conf., ASCE/Cape Town, S. Africa; pp. 1931–1949.

Tauman, J., 1976. Enclosing scheme for bathing-beach development. In Proc. 15th. Coastal Eng. Conf., Honolulu, ASCE; pp. 1425–1438.

Yaacov Nir
Geological Survey,
Israel

Avi Elimelech
Ministry of the Interior,
Israel.

19. Differential response of six beaches at Point Pelee (Ontario) to variable levels of recreational use

Introduction

Since the turn of the century, shoreline erosion has been a major problem observed at Point Pelee National Park, Ontario, Canada. Point Pelee is a cuspate foreland extending six miles into Lake Erie from the north. Both flanks of Point Pelee are noted for their fine sandy beaches (East, 1976). Since the establishment of Point Pelee National Park, a planned pattern of beach usage has evolved with respect to the intensity of visitation by sunbathers and swimmers (East, 1976; Boyd, 1979; Parks Canada, 1982). Since 1918 (Kindle, 1918) a number of attempts have been made to cope with the problems of severe beach erosion around Point Pelee (Kindle, 1933; Persson, 1964; Jarlan, 1966; Kamphuis, 1972; Coakley, 1972; Coakley and Cho, 1972; Coakley et al., 1973; Chrysler and Lathem Consulting Engineers, 1974; East, 1976; Setterington, 1978; LaValle, 1979, 1985, 1986). Prior to 1972, these studies consisted of short term site evaluations and recommendations for the structural protection of key shoreline sites utilizing a variety of armourstone breakwaters, oak-piling breakwaters, concrete tetrapods, gabbion mat systems, as well as a groyne just north of the Northeast Beach, Point Pelee. Since 1973, artificial sediment renourishment has also been attempted at the Northeast Beach with some success (LaValle, 1986). In 1978 Parks Canada, in cooperation with the University of Windsor, initiated an ongoing shoreline monitoring program designed to provide data on dynamics which can be used for conservation planning. One aspect of this project involves the assessment of the impact of the Park's shoreline use and management programs. It is the purpose of this paper, therefore, to examine the relationships between the intensity of beach usage and erosion levels where the effects of lake level fluctuations and wave attack intensity has been held constant. Also an attempt is made to look at the relationships between the intensity of beach usage and erosion levels relative to the erosional response of beach sites dominated by erosion control structures.

The Point Pelee study area

South of the forty-second parallel and extending six miles into Lake Erie, one encounters Point Pelee, the southernmost promontory of Canada, which according to Shaw (1975) is a partially inundated peninsula that was formed 4000 years ago. According to East (1976), Point Pelee is a cuspate foreland flanked by two beach systems enclosing a large marsh area (see Figure 1). East (1976) notes that severe erosional problems exist along the Northeast, East and Southeast beaches, and Shaw (1975) suggests that the eastern beaches are receding at a rate five times the

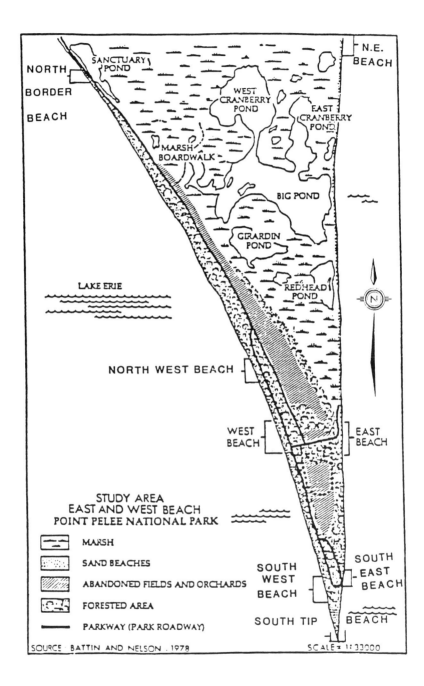

Figure 1.

accretion rate observed along the western beaches. Coakley and Cho (1972) suggest that net shoreline erosion has been taking place since 1947. In response to this situation, a number of shoreline protection schemes have been carried out. At the Northeast Beach, the three hundred metre long northern sector has three lines of concrete tetrapods emplaced along the 1984 shoreline – now they are under water! At the Southwest Beach, there is a combination of a gabbion mat system and an armour-stone breakwater. Also the southern sector of the Northeast Beach was subjected to a single episode of sediment renourishment in 1979.

Both flanks of Point Pelee have sites where bathing is permitted. The East Beach used to be prime bathing area, but erosional activities have reduced the utility of this site. The Southeast Beach near the Point attracts many visitors in a given day – mainly for sunbathing and bird watching. On the western flank, the Northwest Beach is the main bathing beach in the Park, and it attracts more people per square metre than all of the other beaches in the Park. In this study, the Northwest, East and Southeast beaches are classed as intensive recreational use sites. By way of contrast, the North Border Beach in the northwest corner of the Park and the unprotected southern segment of the Northeast Beach represents two relatively unused beaches, because of their remote locations and lack of facilities. In general, the eastern beaches are characterized by finer sediment textures with Phi means ranging between 0.67 and 0.86, but little textural variation exists between them. The western beaches tend to have coarser sediment textures (Shaw, 1975). However, both the Northwest and North Border Beaches have similar Phi means of 0.60 and 0.65. In this study, transects from the relatively unused Northeast Beach will be compared to those of the East and Southeast Beaches in order to evaluate the effects of recreational use impact on the eastern side of the Point. On the western side of the Point, the relatively unused North Border Beach will be compared to the intensely used Northwest Beach. In terms of exposure to wave attack as well as nearshore morphology, these pairings of beach sites provide a high degree of homogeneity with respect to beach orientation, fetch distance, and nearshore configuration.

According to East (1976), the most effective wind generated wave action associated with beach erosion comes from the east, because the fetches on the eastern flank of the Point are much greater than those from the west. In spite of the fact that the prevailing winds are from the southwest, a majority of the erosive storm generated significant waves emanate from the northeast. This contention was verified by a hindcast wave analysis made for the period 1985–1986 at the University of Windsor by Kovacs (1987). Kovacs (1987) hindcasted the wave patterns for Point Pelee on a daily basis using the algorithm developed at the National Water Research Institute in Canada. She then converted these data into estimates of total energy flux using the formula: $E = 1/8 p g H^2 (3600/T)$ where $H =$ wave height, $T =$ wave period, $p =$ water density, and $g =$ acceleration due to gravity. These flux figures were then summated for the eight main points of the compass. The wind data were obtained from the Point Pelee National Park weather station. It was found that 56% of the wave energy flux associated with significant erosion emanated from the northeast during the survey period. It was the northeast winds associated with cyclonic storm activity that was mainly responsible for the higher erosion levels observed on the eastern flank of Point Pelee. For this reason, the eastern beaches were separated from the western beaches for analysis, and one would expect to encounter higher erosion levels along the east flank of the Point.

According to Bukata et al. (1975) there are three main current systems influencing the Point. The two main current systems flow south along each side of Point Pelee, while from the southwest a third surface current flows toward the tip of the Point. The study beaches are mainly influenced by the longshore currents which flow south about two-thirds of the time. However, Shaw (1975) reported that the west side does receive stronger sediment bearing northerly counter flows more frequently than the eastern side, which may also be responsible for the lower net erosion levels experienced along the western beaches.

Within recent years, Lake Erie water levels have risen to record heights which probably has accelerated shoreline erosion processes following the logic of Kindle (1933), Bruun (1962), Rosen (1978), and Hands (1979, 1983). These observers all agree that rising sea levels are a major factor in shoreline retreat. Since the turn of the century, records have been kept describing Lake Erie water levels, and since 1900 there has been a general rise in Lake Erie levels (El Shaarrawi, 1984). Between 1984 and 1986, Lake Erie rose to a record high water level which initiated severe erosional activities all around Point Pelee. Erosion was extremely severe at those sites where structural erosion control devices were emplaced, because many of these structures were designed for much lower water levels, so they failed. Both sides of the Point were subjected to similar water level conditions. For this reason, sites with permanent structural erosion control devices were included in part of the analysis of the differential erosion patterns observed between 1984 and 1986.

Methodology

In order to assess the differential response of the six study beaches to erosional stresses, annual topographic and bathymetric surveys were conducted along standardized survey grids located on each study beach. Land survey was executed using an automatic level along transect lines oriented perpendicular to the grid baseline. Nearshore survey was carried out along these transect lines using a marked cable for horizontal control and a stadia rod to measure the depths to the bottom. This was possible, because the nearshore depths rarely exceeded 3.3 m. Also the nearshore surveys were executed in calm conditions. Once the transects were surveyed, they were plotted using a microcomputer plotting program. This program is designed to plot two successive surveys for each transect, and a subroutine of the program was employed to calculate the net gains or losses of material along the profile line. The losses or gains from the beach portion were then separated from the nearshore gains and losses and integrated into a total for each transect. These were then expressed as the beach net sediment flux for each transect. Also changes in shoreline position were surveyed for each transect by determining the net difference in the distance of the shoreline from the baseline for successive surveys. Both net shoreline positional change and beach net sediment flux were then used as criterion variables to assess the differential erosional patterns observed along Point Pelee.

Table 1. Beach net sediment flux variations between high density use and low intensity use sites for eastern and western Point Pelee: A two way analysis of variance.

Source	Sum of squares	d.f.	Mean square	F-ratio
Between East and West Beaches	1312	1	1312	11.69*
Between use intensities	562	1	562	5.01*
Interaction effects	135	1	135	1.20
Error	1796	16	112	
Total	3805		200	

Cochran's $C = 0.48$ (not significant)   * = significant at 0.05 level.

	Data means		
	Low intensity use	High intensity use	Marginal mean
West side	− 3.0 a	+2.4	− 0.3
East side	− 24.4 ab	− 8.6 b	− 16.5
Column means	− 13.7	− 3.1	− 8.4

$D_{crit.}$ for Newman-Keuls test (at the 0.5 level) = 14.2

a: Difference between East and West side low intensity use sites exceeds 14.2 and is significant.
b: Difference between high and low intensity use sites is greater than 14.2 and is significant along the East side of the Point.

Using shoreline positional change and beach net sediment flux as criterion variables, the differential response of the intensively used and relatively unused beach sites were compared using a two way analysis of variance design where one dimension of the model was used to control the differences in expected erosion levels for the eastern as opposed to the western beaches. Each beach site will be represented by five transect measurements. Cochran C tests were employed to assess the homogeneity of variance assumptions. Following the assessment of the differential response of intensively used versus relatively unused beach site responses to erosional stresses, the responses of beaches dominated by erosion control structures – concrete tetrapods, gabbion mats, or rock breakwaters were to be assessed by running another two way analysis of variance with the structurally protected beaches being a third category in the intensity of use classification. In both sets of analysis of variance tests, Newman-Keuls tests were employed to assess individual differences where significant F ratios were observed.

Results

Between 1984 and 1986, Lake Erie rose to record heights generating widespread

Table 2. Net shoreline change variations between high intensity use and low intensity use sites for eastern and western Point Pelee: A two way analysis of variance.

Source	Sum of squares	d.f.	Mean square	F-ratio
Between East and West Beaches	274	1	274	8.66*
Between use intensities	72	1	72	2.28
Interaction effects	51	1	51	1.62
Error	506	16	32	
Total	903	19	48	

Cochran's $C = 0.43$ (not significant)   * = significant at the 0.05 level.

	Data means		
	Low intensity use	High intensity use	Marginal mean
West side	− 1.8 a	− 1.2	− 1.5
East side	− 12.4 a	− 5.4	− 8.9
Column means	− 7.1	− 3.3	− 5.2

$D_{crit.}$ for Newman-Keuls test (at the 0.05 level) = 7.6.

a: Difference between East and West low intensity use sites exceeds 7.6 and is significant.

accelerated erosion throughout the Point Pelee Beach system. During this episode of high water levels, Point Pelee's shorelines retreated an average of 5.2 m along sites not influenced by structural erosion control devices, but along those Northeast Beach sites fronted by three lines of concrete tetrapods, the mean level of shoreline retreat was 90 m. Those Southwest Beach sites underlain by the gabbion mat system experienced shoreline retreat rates averaging nearly 6.8 m. In fact, the entire gabbion mat network has been destroyed. It is apparent that the sites where structural protection was emplaced were afflicted with poorly designed structures based on an underestimation of the magnitude of the record water level conditions. However, hindsight is always 20/20, so one must not be too critical. Beach erosion rates averaged 8.4 cubic metres per metre of beach front at the unprotected sites but beach erosion levels at the Northeast Beach concrete tetrapod site averaged 53 cubic metres per metre of beach front.

When the two way analysis of variance was performed on the high intensity use versus low intensity use unprotected beaches using beach net sediment flux as a criterion variable, significant variations in erosion levels between the east and west coast beaches was observed ($F = 11.69$, see Table 1). This supports the contention that erosion levels tend to be higher on the eastern side of the Point. A significant difference was observed in beach net sediment flux levels between the high intensity use and low intensity use beaches ($F = 5.01$), but an examination of the

Table 3. Two way analysis of variance between high intensity use, low intensity use, and protected sites for eastern and western beach net sediment flux data.

Source	Sum of squares	d.f.	Mean square	F-ratio
Between East and West Beaches	4839	1	4839	9.09*
Between use intensities	3997	2	1999	3.73*
Interaction effects	1404	2	702	1.32
Error	12774	24	532	
Total	23015	29	794	

Cochran's $C = 0.83$ * = significant at the 0.05 level.

	Data means			
	Low intensity use	High intensity use	Protected	Marginal X
West side	− 3.0	+ 2.4	− 9.2g a	− 3.3
East side	− 24.4	− 8.6 b	− 53.0c ab	− 28.7
Column means	− 13.7	− 3.1	− 31.1	− 16.0

$D_{crit.}$ for Newman-Keuls test (at the 0.05 level, $r=2$) = 30.1.
$D_{crit.}$ for Newman-Keuls test (at the 0.05 level, $r=3$) = 36.4.
a: Difference between East and West protected sites exceeds 30.1 and is significant.
b: Difference between East side high intensity use site and East side protected site (concrete tetrapods) exceeds 36.4 and is significant.
g – Protected by gabbion mats.
C – Protected by concrete tetrapods.

Newman-Keuls results suggest that this is only found along the eastern side of the Point where the relatively unused portion of the Northeast Beach experienced significantly higher erosion levels than the heavily used East and Southeast Beaches. This may be due to the fact that the Northeast Beach has a less effective vegetation cover and may be influenced by the high erosion levels associated with the adjacent concrete cross system found in the northern half of the Northeast Beach.

When the two way analysis of variance was performed on the high intensity use versus the low intensity use unprotected beaches using net shoreline change as a criterion variable, significant variations in shoreline retreat levels were observed between the east coast and west coast beaches. (F = 8.66, see Table 2). No significant differences were noted between the high intensity use and low intensity use beaches with respect to shoreline retreat levels. On the average the eastern beaches retreated 8.9 m while the western beaches retreated an average of 1.5 m.

When two way analysis of variance was executed on the beach net sediment flux data derived from high intensity use, low intensity use and structurally protected

Table 4. Two way analysis of variance between low intensity use, high intensity use and protected sites for eastern and western Point Pelee net shoreline change data.

Source	Sum of squares	d.f.	Mean square	F-ratio
Between East and West Beaches	8036	1	8036	19.52*
Between use intensities	12571	2	6286	15.27*
Interaction effects	9678	2	4839	11.75*
Error	9880	24	412	
Total	40165	29	1385	

Cochran's $C = 0.93$ * = significant at 0.05 level.

	Data means			
	Low intensity use	High intensity use	Protected	Marginal X
West side	− 1.8	− 1.2	− 6.8 g	− 3.3
East side	− 12.4	− 5.4	− 90.2c	− 36.0
Column means	− 7.1	− 3.3	− 48.5	− 19.6

g − Protected by gabbion mats.
c − Protected by concrete tetrapods.

site data, significant variation in beach sediment flux averages was noted between the eastern and western beaches ($F = 9.09$, see Table 3), but in this situation the main source of significant variation came from the concrete tetrapods portion of the Northeast Beach which lost an average of 53 cubic metres per metre frontage of beach sediment. When this is compared to an average loss of 9.2 cubic metres per metre frontage from the gabbion mat section of the Southwest Beach, it represents a loss that is over five times that observed at a protected western beach. Also a significant level of beach net sediment flux variation existed between the high intensity use, low intensity and structurally protected sites, but a Newman-Keuls analysis of the data led to the inference that the main source of significant variation involved the differences between the concrete cross protected Northeast Beach and the high intensity use East and Southeast Beaches. In this analysis, significant heteroschedacity of variance was observed ($C = 0.83$, significant at the 0.05 level), so the results of the analysis of variance are at best only approximate. The main source of this heteroschedacity of variance was traced to the concrete tetrapod area of the Northeast Beach.

When two way analysis of variance was executed on net shoreline positional change data for the high intensity use, low intensity use and structurally protected site data, significant interaction effects were observed between the dimension representing beach use and the dimension separating the more heavily stressed beaches from the western beaches ($F = 11.8$, see Table 4).

Also significant heteroschedacity of variance was observed in this analysis of variance as evidenced by a Cochran's $C = 0.93$. The main source of these problems is located in the concrete tetrapod section of the Northeast Beach which experienced extraordinarily high erosion levels between 1984 and 1986. However, the presence of significant interaction effects and heteroschedacity of variance effects has rendered this phase of the investigation inconclusive.

Summary and conclusions

An attempt was made to assess the impact of beach recreational use intensity on beach erosion and shoreline retreat levels where the relative effects of wave attack intensity are controlled. Also an attempt was made to assess the comparative response of sites dominated by structural erosion control in another phase of the study. Based on hindcast wave data, it was found that 56% of the total energy flux associated with significant wave heights emanated from the northeast during 1984–1986. Therefore the eastern beaches were expected to experience higher erosion levels, and this was the case with respect to both beach net sediment flux and net shoreline positional change. When the high intensity use beach sites were compared to low intensity use beach sites for variations in beach net sediment flux, the only significant difference was found between the eastern high intensity use East-Southeast Beaches and the low intensity use Northeast Beach, which may be associated with a less stable vegetation community encountered at the Northeast Beach or the effects of the adjacent concrete tetrapod system on the other part of the Northeast Beach. However, it seems plausible to infer that the intensity of beach use for swimming and sunbathing has not had a significant detrimental effect on beach erosion rates. While shoreline retreat levels were significantly higher on the eastern side of the point, no significant variations were observed between high intensity use sites and low intensity use sites. When sites protected by structural control devices were included in the analyses of variance models, they tended to reflect significantly higher erosion levels, but these results were afflicted with heteroschedacity of variance problems associated with the total failure of the concrete tetrapod system. Also when variations in net shoreline retreat levels were assessed in analysis of variance models including sites with structural protection, significant interaction effects rendered the results inconclusive.

Based on the results of this investigation, it could be inferred that the intensity of beach visitation and use has little effect on shoreline retreat and beach net sediment flux levels. However, the use of improperly designed structural erosion control systems may be a much greater threat to a beach system, so it may be better to allow people to use the beach for recreation without trying to control the beach's behaviour through the implementation of structural control devices.

Cited references

Bishop, C. T. and Donelan, M. A. (1985). Wave hindcasting, National Water Research Institute, Burlington, Ont.

Bruun, P. (1962). Sea level rise as a cause of shore erosion. *Am. Soc. C.E. Proc. Jour. Waterways and Harbours Div.* **88**, 117–130.

Bukata, R. P., Haras, W. S. and Bruton, J. E. (1974). The application of ERTS–1 digital data to water transport phenomena in the Point Pelee-Rondeau area. Proc. 19th Cong. Int. Assoc. of Limnology, pp. 168–178.

Boyd, G. L. (1979). Review of literature concerning erosion and accretion of the Canadian shoreline of the Great Lakes. Unpub. manuscript, Ocean and Aquatic Sciences CCIW, 24 pp.

Chrysler and Lathem Consulting Engineers (1974). Northeast beach erosion study, Point Pelee National Park. Report to the Dept. of Indian Affairs and Northern Development, 28 pp.

Coakley, J. P. (1972). Nearshore sediment studies in western Lake Erie. Proc. 15th Conf. Great Lakes Res. IAGLR, pp. 330–343.

Coakley, J. P. (1976). The formation and evolution of Point Pelee, Western Lake Erie. *Can. J. Earth Sci.* **13** (1), 136–144.

Coakley, J. P. and Cho, H. K. (1972). Shore erosion in western Lake Erie. Proc. 15th Conf. Great Lakes Res. IAGLR, pp. 344–360.

Coakley, J. P., Haras, W. S. and Freeman, N. G. (1973). The effect of a storm surge on beach erosion, Point Pelee. Proc. 16th Conf. Great Lakes Res. IAGLR, pp. 377–389.

East, K. (1976). Shoreline erosion: Point Pelee National Park: a history and policy analysis. Ottawa: Parks Canada. 66 pp.

El Shaarawi, A. H. (1984). Statistical assessment of the great Lakes surveillance program 1966–1981, Lake Erie. C.C.W.I. Science Series #136.

Hands, E. B. (1979). Changes in rates of shore retreat, Lake Michigan 1967–1976. U.S. Corps of Eng., CERC Tech. Paper 79–4, 71 pp.

Hands, E. B. (1983). The Great Lakes as a test model for profile response to sea level changes. Handbook of coastal processes. Florida: CRC Press, pp. 167–189.

Jarlan, G. E. (1966). Memorandum concerning preliminary recommendations for the protection of shoreline against erosion at Point Pelee National Park, Ontario, June 28, 1966. National Parks Research Collection, 17 pp.

Kamphuis, J. W. (1972). Beach data collection programme in Canada. Proc. Port, and Ocean Eng. Under Arctic Conditions, Tech. University of Norway, pp. 1186–1202.

Kindle, E. M. (1918). Letter to Mr. William McInnis, Directing Geologist Can. Geol. Surv., Dept. of Mines (Public Works), reporting on erosion conditions at Point Pelee, Aug. 1, 1918.

Kindle, E. M. (1933). Erosion and sedimentation at Point Pelee. Dept. of Mines 42nd Annual Report, Part 2, 22 pp.

Kovacs, M. A. (1987). Spatial response to shoreline processes: South Tip Beach, Point Pelee. Unpub. B.A. Thesis, University of Windsor, 195 pp.

LaValle, P. D. (1979). Shoreline erosion control program at the Northeast Beach, Point Pelee in Resource Allocation Issues in the Coastal Environment. Arlington: The Coastal Society, pp. 289–305.

LaValle, P. D. (1979). Monitoring shoreline erosion at the Northeast Beach, Point Pelee. Windsor Essex County field guide. Dept. of Geography, University of Windsor, pp. 78–88.

LaValle, P. D. (1985). Shoreline survey of eight beaches along Point Pelee. Unpub. report to Parks Canada, 62 pp.

LaValle, P. D. (1986). Northeast Beach Survey: May, 1986. Unpub. report to Parks Canada, 62 pp.

Parks Canada (1982). Point Pelee National park management plan. Cornwall: Ontario Region, Dept. of Indian and Northern Affairs, 146 pp.
Persson, R. H. (1964). Memorandum concerning shore protection at Point Pelee. Ottawa: Dept. of Public Works.
Rosen, P. S. (1978). A regional test of the Bruun rule on shoreline erosion. *Marine Geol.* 26, 7–16.
Setterington, W. J. (1978). Erosion survey at the Point Pelee National Park. Unpub. report to Parks Canada.
Shaw, J. R. (1975). Coastal response at Point Pelee, Lake Erie. Unpub. Ocean and Aquatic Sciences CCIW, 81 pp.

Consulted references

Bayly, I. L. (1977). A rationale for repair and rehabilitation of Point Pelee National Park. Ottawa: Carleton University, 111 pp.
Boulden, R. S. (1975). Canada/Ontario Great Lakes shore damage survey technical report. Ottawa: Environment Canada and Ministry of Natural Resources, 97 pp.
LaValle, P. D., Rudakis, P. and Khan, B. (1979). Shoreline dynamics at the Northeast Beach, Point Pelee. Unpub. report to Parks Canada, 175 pp.
LaValle, P. D. (1980). Shoreline variations at the Northeast Beach, Point Pelee. Unpub. report to Parks Canada, 119 pp.
LaValle, P. D. (1981). East Beach and Northeast Beach shoreline patterns, May-November, 1981. Unpub. report to Parks Canada, 47 pp.
LaValle, P. D. (1981). Shoreline variations at the East and Northeast Beaches, Point Pelee. Unpub. report to Parks Canada, 56 pp.
LaValle, P. D. (1982). Shoreline variations at the East and Northeast Beaches, Point Pelee. Unpub. report to Parks Canada, 53 pp.
LaValle, P. D. (1983). Northeast and West Beach shoreline variations at Point Pelee. Unpub. report to Parks Canada, 41 pp.
LaValle, P. D. (1984). Shoreline survey of eight beaches along Point Pelee. Unpub. report to Parks Canada, 61 pp.
LaValle, P. D. (1985). Coastal management of the North Shore of Lake Erie. AAG Field Trip Guide 1985, pp. 55–68.
Ontario Ministry of Natural Resources (1981). Great Lakes shore processes and shore protection. Toronto: OMNR 77 pp.
Saville, T. (1953). Wave and lake level statistics for Lake Erie. U.S. Army Corp. of Eng., Beach Erosion Board Tech. Memo. 37, 14 pp.
Skafel, M. G. (1975). Longshore sediment transport at Point Pelee. Unpub. report to CCIW.
Turnbull, J. E. (1963). Memorandum concerning shoreline erosion at Point Pelee. Unpub. Parks Canada memo.

Placido D. La Valle
Department of Geography,
University of Windsor,
Canada

20. Anthropogenic effects on recreational beaches

Anthropogenic effects on the sea beach are traditionally regarded in terms of construction and economic activities. Such an approach tends to oppose nature to society as a whole, or large groups thereof. Meanwhile, interaction between nature and small groups of people or even individuals is of great interest.

As applied to beaches, this problem is most evident in the recreational zones, where impact of holiday-makers is most pronounced. The great importance of this factor in shore evolution was first mentioned by W. Bülow: 'Holiday-making causes as much damage to the shore, as an average storm surge' (Hupfer, 1982).

Shattered dirty beaches can produce rather an unfavourable impression. Putting emotions aside, it should be noted that the main reason why the man-made traces are clearly visible is an inadequate organization of recreational zones. Thus, beach planning (selection of certain sites for parking, camping, garbage disposal, etc.), cleaning and sanitary treatment can eliminate many negative effects of recreational activities.

This problem has a number of more complex aspects which should be taken into account in the planning of recreational zones. The first of them is mechanical action of men on beaches and beach sediments. As men and cars move around, the whole of the beach surface or its separate areas change with beach sediment shifting towards the water line, the shoreline being deformed, sediment ground.

Some aspects of this complex phenomenon were studied on sandy-shelly sediments of the Azov sea accumulative forms in 1985–1986. Studies of anthropogenic beach deformation were carried out on sites marked by control points using successive levelling and theodolite surveying techniques. The nature of shell sediment transformation was studied on experimental sites in conditions close to natural.

Figure 1 shows changes of the beach surface relief on the eastern side of Dolgaya spit, normal to the shoreline, in a 10 m zone along the water line. It can be seen that after weeks' stay of 25 people the surface of comparatively gently dipping beach (gradient tgd = 0.16) composed by quartzose and detrital sand (Md = 1.2 mm) was transformed. Reliability of the anthropogenic component of the beach deformation was substantiated by the reworking and levelling of the beach up to the foredune by the previous storm.

Holiday-makers cause mound and trough formation to appear, which finally leads to considerable sand shifting towards the water line. Local shifting of sand material reached 0.02–0.2–0.3 m^3 per running metre. It should be noted that the erosive scarp shown on Figure 1, was formed due to anthropogenic shifting of beach sediment (0.13 m^3 per running metre) towards the water line. As a result, even mild swash (wind velocities up to 2–3 m s^{-1} caused additional shifting of 0.05–0.10 m^3 of sand from each running metre towards waterline zone. Assuming

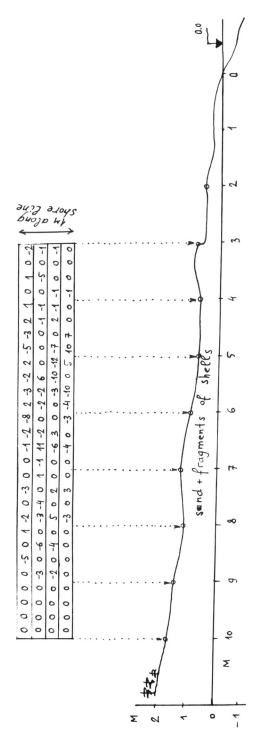

Figure 1. Deformation area, cm, of beach surface of Dolgaya spit, after stay of 25 men for a week.

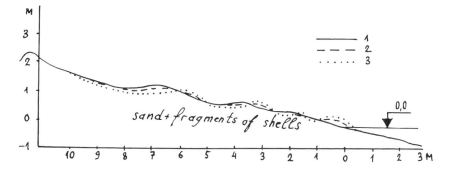

Figure 2. Deformation of a beach cross-sectional profile on the eastern side of Dolgaya spit due to anthropogenic impact. 1 – natural slope, 2 – beach after twentyfold crossing by 2 men up and down the beach, 3 – the same, after fortyfold crossing by 4 men.

the length of the saturation zone of sediment flow branch to be 100 m, abnormal input to it amounts to 5–10 m^3. The value is not very high, but on beaches deficient in sediment such anthropogenic impact can lead to a slow shore erosion.

Figure 2 shows sand material shifting across the beach profile composed by sandy-shelly sediment. The changes shown were caused by two persons who had walked up and down the beach 10 times each in both directions followed by 4 persons, who made 20 trips in the same directions. The most pronounced changes took place in the lower profile near the storm bars composed mostly by *Cerastoderma lamarcki* shells (up to 70%). In this part sediment compaction and shifting was maximal. An 'anthropogenic' bar of 2 m^3 per running metre was formed along the water line.

An extremely negative effect can be caused by anthropogenic reforming of the upper profile. Using the upper parts of beach bars and spits for passage of cars and camping inevitably causes troughs and pit formation which can lead to wave rolling over the hollows and breaching of the accumulative forms.

Table 1. Sphere ultimate stress causing shell destruction in the Azov shell-fish on a steel plate.

Shell-fish species	Shell diameter (mm)	Shell thickness (mm)	Shell-fish age	Ultimate stress on shell sphere (kg-force per cm^2)
Cardium edule	25.0–30.0	2.0–3.0	Karangatsk	17.0–19.0
Cerastoderma lamarcki	20.0–25.0	0.7–0.9	Recent	4.0–5.0
Cerastoderma lamarcki	25.0–30.0	0.5	Recent	0.7–1.0
Monodacna colorata	30.0	0.4	Recent	0.5
Chione gallina	27.0	0.9	New-Azov	1.79

The second important aspect of this problem is the change of beach sediment grain-size. It is of particular significance for beaches composed of shell material. It is illustrated by the results of shell strength test, carried out on a special testing unit UM–10 TM (Table 1).

Table 1 shows that many recent shells, especially freshwater ones, are rather brittle. Taking into account an average stress of up to 0.9–1.2 kg force per sq cm developed by an individual walking on loose sand, the impact on shelly beach sediments becomes clear. Our tests show that after a single step of a shod foot of a man weighing 80 kg, 4 to 5 shells of a thin-walled *Cerastoderma lamarcki* with shell diameter 11–25 mm are crushed out of each 25 shells. They are broken into particles with a grain size from 1 cm to 3–5 mm. Where the beach is covered by two shell layers, 5–7 shells are crushed. As a result the median diameter of beach sediments decreases from 12–15 mm to 10–11 mm.

Other shells – *Mya arenaria, Mytilus galloprovincialis, Anadonta tumidus, Monodacna colorata, Dreissena polymorpha, Donacilla barnea*, etc., resting on sand are crushed even when stepped upon by a bare foot. If they rest on shell sublayer, the process is even more intensive.

The general assumptions show that transfer of beach material from one grain-size range to another affects its dynamics. Thus, a relatively large part of sediment will not return to the beach after washout has finished. As a result the pre-storm beach slope under natural conditions and the post-storm beach slope after the anthropogenic change of beach sediment can differ. Estimates based on T. Sunamura (1980) relation, give the slope change of 15–22%.

The data on man's impact on beaches apply mainly to sandy-shelly accumulative forms of the Azov sea. Moreover, observations on the Black sea coast showed that even pebble beaches suffer from holiday-makers. There is no doubt that holiday-makers contribute to the rounding of beach material, especially that composed of soft rocks, such as silts, marls, siltstones etc. Shore profile is also affected and it is still unknown whichever suffers most – sandy or pebble shore. It is suggested that gravitational shiftings initiated by man, can be more significant in pebble beaches, although exact data are not available.

It can be concluded that a man interacts with a beach not only indirectly or through machines. Individual holiday-makers may bring unintentional changes to the beach profile. This fact should be taken into consideration when planning measures for recultivation, feeding up of beaches. Without proper control beach degradation may result from massed stress of holiday-makers.

Reference

Hupfer, P. 1982, The Baltic Sea – Small sea, Great problems, Leningrad, Hydrometeoizdat, 135 pp.

Yurii V. Artukhin
Geological-Geographical Department,
Rostov State University,
U.S.S.R.

21. Formulating policies using visitor perceptions of Biscayne National Park and seashore

The site

This study was conducted in a national park located near Miami, Florida. As Figure 1 depicts, most of the park is reef and water, but within its boundaries there are many keys or islands forming a north-south chain bounded by Biscayne Bay on the west and the Atlantic Ocean on the east. The area was selected for demonstration of the evaluation approach, discussed below, because it is part of a large diverse recreation system that includes public parks, beaches, marinas, boating, diving, and commercial water-related activities. It is a unique recreation area annually serving millions of visitors both domestic and foreign.

Evaluation assumptions

The approach to studying public reaction to the Bay uses objective data collection methods to obtain information on specific types of users, or reference public's, ratings of facilities, programs, and services. These data were used to assess fulfillment of Park goals, objectives, and standards, which then form the basis for recommendations that seek to improve management efficiency or maximize overall public gain. It is unlike other management or planning methods in that it blends attitudinal issues such as satisfaction, environmental concern, and expectations relevant for behavior in the Park situation into the evaluation, planning, and policy formulation process. The strategy borrows generously from a variety of disciplines in constructing the approach. Its strength lies in objectively questioning the social acceptability of what Park management has previously accomplished.

Evaluation goals

Differing assessments and perceptions of the services, facilities, and programs offered at Biscayne National Park were examined. The empirical research was guided by the policy evaluation objectives established by Bauer and Gergen (1986): 1) to indicate realistic limitations on the type of Park among potentially competing groups of recreationists; 2) to supply that understanding of current practices and existing forms of organization which will facilitate the introduction of better procedures; and 3) to improve those aspects of the process of policy formation which fall outside the framework of formal decision theory. Given these broad goals, our survey evaluated program implementation by the public throughout the course of a year. It expands the scope of the traditional policy and management

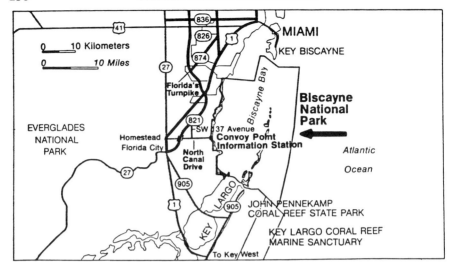

Figure 1. Location of Biscayne National Park.

decision-making process by including public perceptions. These included measures of satisfaction with existing facilities, services and programs, as well as an evaluation of realistic management options for change within the Park. Expectations and aspirations for improving park functioning are thus included in the management and policy determination process. Thus the evaluation was also designed to provide data which reach beyond the present circumstances in Biscayne Bay.

Evaluation approach

The focus of data gathering was first narrowed to identifying those visitor benefits and issues that site managers could realistically change within budgetary and personnel constraints. The evaluation team stressed that managers were partners in the design of the evaluation since the research focused on issues and problems under direct management control. Data gathering centered on the possibility of reform within existing programs and hence perceptions of conflict and factious conditions were sought from staff during the development of the Park's evaluation instrument. Locating conflict among recreationists, park management, and the marine resource helped to structure the issues addressed by the evaluation protocol. Data were specifically gathered to evaluate the publics response to Biscayne's physical facilities, support services, recreation programs, and park specific problems. Demographic information profiling the Bay recreationists was also collected and analyzed in relation to visitor expectations and their perception of current conditions. These demographic variables included age, sex, ethnicity, education and socioeconomic factors.

After constructing the questionnaire, fieldwork methods were developed to disseminate the instrument to two samples. The first sample consisted of randomly selected visitors to the park during the summer and winter. The second sample consisted of randomly selected persons with boats registered in Dade County Florida. The first sample of visitors were intercepted in the park and given a questionnaire to return after their visit. The second sample received their questionnaire in the mail and were requested them to return it by mail. A total of 1760 forms were distributed, with samples of 440 intercept and 440 direct mail distributed in the summer, and identical samples in the fall. This seasonal weighting was developed using 1985 visitor statistics, which placed approximately half of the Park's visits during the summer. Once disseminated, 69.2% of the intercept and 49.3% of the direct mail questionnaires were returned.

At each of the sites 'rated' by the questionnaire, every fifth boat, or recreation vehicle that entered or passed a central point was considered eligible for inclusion in the sample. This technique was chosen to ensure an unbiased random sample.

Overall, 59.2% of the forms distributed were returned. This completion rate was secured through the application of the Dillman technique (*Mail and Telephone Surveys*, New York: Wiley, 1984) for maximizing return rates from mailback surveys. This technique includes the use of 1) a personal letter-reminder mailed one week after distribution of a questionnaire, 2) a personal letter and supplemental questionnaire sent to those who had not replied in three weeks after the original distribution, and 3) a final personal letter and questionnaire sent to all who had not returned the questionnaire seven weeks after distribution.

Presentation and analysis of the evaluation data

Frequency distributions and descriptive statistics were calculated for each question. Demographic profiles of typical park visitors, their travel party, and trip characteristics were developed to describe who was using the park. The items evaluating specific park sites, services, and programs were also summarized. These data were then classified by ethnicity of the visitor (Hispanic, Anglo), season (summer or winter), and type of survey administration (intercept or direct mail) to determine if significant differences could be found across the seasons, by ethnicity, and by type of data collection. These tabulations were supplemented with classifications by type of park use in conjunction with the demographic variables.

Because of the large number of questions included in the survey and the huge number of tables that resulted when the data were classified by type of data collection, a decision was made to organize the analysis around the structure of the original questionnaire booklet. The survey instrument thus became a template that displays the frequency distributions of the entire sample of 1043 travel parties. This structure was chosen because the responses to a vast majority of the questions, such as those rating facilities or incidence of problems, displayed little variation across seasons or methods of data collection. In the interest of brevity, the summary distributions discussed below were reported to park management, and any significant differences across levels of the above variables are discussed in the text. Chi-square analyses or t-tests were used to determine statistical significance of differences (alpha = 0.05 or less for the appropriate test statistic).

Table 1. Part I Awareness and visitation of Biscayne National Park.

1.	Were you aware that this area is a national park?			
		83.3% YES	16.7% NO	$n = 1043$
2.	Have you ever visited or passed through this area?			
		91.9% YES	8.1% NO	$n = 1043$
3.	Have you ever visited this area for pleasure or recreational purposes?			
		89.2% YES	10.8% NO	$n = 1043$
4.	In the past 3 months, about how many times have you visited this area for pleasure or recreational purposes?			
	Number of visits:		Average 6.2	$n = 930$
5.	In the past 6 months, about how many times have you visited this area for pleasure or recreational purposes?			
	Number of visits:		Average 12.0	$n = 930$.

Evaluation results

Figure 2 and Table 1 were complimentary pages that appeared facing each other in the survey questionnaire (Table 2). Respondents were instructed to indicate whether they were aware that the area represented by the Park on the map was a national park, if they had ever visited or passed through the area, and if they had ever used the Park for recreation purposes. If the respondent had visited this area (n=930) they were then asked to complete a series of items concerning their previous visit(s) and impressions of the Park. If those contacted by direct mail had never visited the Park they were instructed to complete items in the latter half of the questionnaire (Table 2) and skip the items directly related to evaluation of Biscayne. This structure allowed comparisons of the demographics and attitudes of visitors to nonvisitors.

After determining how often people visited the park, the next section of the questionnaire, summarized in Tables 3 and 4, requested that visitors detail information about the characteristics of their last visit, travel party, and motivations for using the park's facilities. These data were used by park management to form a prototype of the typical Biscayne visit and travel party.

After supplying this basic information about their last visit, respondents were then asked to detail the sites that they had used or observed on their last visit. These sites and facilities were identified by Figure 2 and the grid in Table 5. If the visitor had used or observed a site or facility, he was then requested to evaluate it using an excellent 1) to poor 4) four-point scale. From a policy perspective, these data indicate relative satisfaction with the park's current configuration. These conclusions were reinforced by data from the next section of the questionnaire (Table 6) that sought more general evaluation of services, facilities, and programs within the park. The distributions on these items were again skewed to the positive. However,

Figure 2. Biscayne National Park.

Table 2. Duration and boat use on respondent's host visit to Biscayne National Park.

6.	On *your last visit* to this area, did you stay overnight in the Biscayne National Park?			
	n=930		Yes-Overnight Hispanic	Non-Hispanic
	80.4% NO	19.6% YES	12.5%	20.9%
6.bis.	Was your overnight stay on a boat?			
	8.2% YES		91.8% NO	
7.	How many hours did you spend in the Park during your last vist?			
	Number of hours:		10.0 average	
8.	On your last visit to this area, did you use a boat?			
	5.0% NO	95.0% YES	n = 923	

If yes, please describe the boat (sail, motor, length, etc.): n=856
canoe, rowboat 0.7 houseboat 0.1
motorboat LT 20' 36.1 sailboat LT 20' 1.8
motorboat 20–35' 45.3 sailboat 20–30' 7.1
motorboat 36'+ 4.6 sailboat 36'+ 4.3

Table 3. Travel party characteristics and reasons for visiting Biscayne National Park.

9.	Counting yourself, how many people were with you during your last visit to the Biscayne National Park?
	Number of people: 4.3 average
10.	Which of the following *best* describes the group of people you were with?

Check one	Description	
3.9	Alone	n = 930
20.4	A couple	
11.3	Family with children age 12 and under	
5.9	Family with children over age 12	
5.3	Family with children 12 and under and over 12	
14.9	Special interest group (swimmers, fishermen, campers, church group, etc.)	
11.2	Two or more families or relatives together	
27.1	Two or more friends together	

11. What were the major reasons for your last visit to the Biscayne National Park?

Check reasons that apply	Reasons for trip	
23.8	To look at scenery	$n = 930$
7.3	To visit park facilities, such as visitor centers, demonstration areas, etc.	
84.0	To participate in outdoor recreation activities, such as swimming, fishing, diving, etc.	
9.8	Just passing through to another location	
10.9	Other	

these distributions also illustrate a small, but significant minority of travel parties that hold negative opinions about some aspects of the Park. The effect of these opinions was evident in response to Section IV of the instrument concerning problems within the Park (Table 7), suggestions for change in the park (Section V, Table 8), and interest in activities available within the park (Section VI, Table 9). This section of the questionnaire proved most useful in detailing the kinds of policies different publics would prefer the park management to pursue. When analyzed by ethnicity, Tables 7 through 9 illustrated that Hispanic patrons of the park have much different expectations and desires for their recreation within the park than those of non-Hispanic residents. Given Miami's large Hispanic population, these differences were explored in detail. Hispanics were found to desire a more developed recreation area with organized services, rather than a less developed, more wilderness recreation area preferred by non-Hispanics. Significantly higher rates of Hispanics want the park to change its facilities and services such that there are more signs, control and patrolling by rangers, picnic areas, live demonstrations, mosquito control, scuba areas, navigational aids, developed camp sites, mooring areas, and food/fuel availability. Hispanics report higher rates of problems over lack of written rules, children's activities, insects, wildlife, and dirty restrooms than were manifest by non-Hispanics. Apparently, as a group, Hispanics bring a different set of expectations about what they would like to find within this park than those expressed by non-Hispanic patrons. Hispanic recreationists preferred a more structured setting with more formally organized activities, while Anglos appeared to want the opposite.

The remainder of the questionnaire concerned visitor and nonvisitor preferences for outdoor recreation (Table 10), attitudes toward environmental use and regulation (Table 11), and demographics. From a policy perspective, the environmental use items again found significant differences between Hispanics and Anglos. More Hispanics favored intrusive use of the environment by man, even if such use would ultimately abuse the environment. These data suggest that attempts to maintain the marine resource as a natural area with minimal impacts by man may encounter resistance among a major group of persons currently using the Park.

Table 4. Part II Evaluation of sites and facilities.

On the opposite page is a map of Biscayne National Park. Using this map for reference, please identify those areas or sites within the Park which you visited on *your last* trip to the Park by placing a check in the column to the left of the sites named. Then, please rate your overall impressions of each site you visited on your last trip to the Park by placing a check to the right of each site in the appropriate box (excellent, good, fair, or poor) in the 'Ratings of sites and facilities used' portion of the page. Review the facilities and services listed under each of the sites you visited and place check in the first column to the right of each facility or service (Picnic Area, Dock, etc.) that you used or observed. Finally, rate each facility or service, you used or observed under the 'Ratings of sites and facilities used' portion of the page.

% Visited this site (*)	Sites Facilities	% Used or observed facility (*)	Ratings of sites and facilities used(*)				Average* rating
			Excellent	Good	Fair	Poor	
22.4	Boca Chita Key		38.7	37.6	18.3	5.4	1.89
	Harbor	20.0	42.0	43.7	9.8	4.6	1.77[bc]
	Picnic area	14.0	29.0	39.5	23.4	8.1	2.09[d]
	Campground	07.0	19.7	32.8	27.9	19.7	2.41
	Trails	09.0	26.2	35.0	25.0	13.7	2.25
13.3	Convoy Point		24.2	57.6	9.1	9.1	2.06
	Glass-bottom boat	1.0	22.2	66.7	-	11.1	2.00
	Snorkeling tour	1.0	18.2	45.5	27.3	9.1	2.30[d]
	Trail picnic area	3.3	25.0	57.1	10.7	7.1	2.00
	Ranger-led activities	1.3	23.1	46.2	-	30.8	2.50
	Restrooms	8.9	13.4	54.9	18.3	13.4	2.34
	Parking	11.0	28.0	49.0	11.0	12.0	2.08
51.8	Elliot Key		52.3	37.6	8.3	1.8	1.60
	Dock	44.8	52.5	37.6	8.4	1.4	1.59
	Picnic area	34.1	31.7	50.8	13.8	3.8	1.90
	Visitor center	29.8	37.4	43.6	14.3	4.8	1.87[b]
	Trails	24.5	26.7	45.8	23.1	4.4	2.05[b]
	Campground	17.7	19.0	52.8	19.6	8.6	2.19
8.5	Adams Key		47.1	29.4	23.5	-	1.76
	Dock	7.2	41.8	35.8	22.4	-	1.82
	Nature trail	2.7	24.0	28.0	40.0	8.0	2.33
	Picnic area	4.1	20.5	43.6	33.3	2.6	2.18[b]
	Restrooms	4.8	16.7	40.5	23.8	19.0	2.45
41.0	On the Reefs		61.1	30.6	6.9	1.4	1.5
	Mooring/Anchorage	27.6	32.9	43.4	15.1	8.5	2.01
	Markings/Navigational aids	33.9	37.7	38.4	22.3	6.6	2.03
	Safety patrol	20.1	27.1	43.6	15.4	13.8	2.18
43.4	On the Bay		41.5	45.7	11.7	1.1	1.71

Table 4 (continued)

Sites Facilities	Visited this site (*) %	Used or observed facility (*) %	Ratings of sites and facilities used(*)				Average* rating
			Excellent	Good	Fair	Poor	
Mooring/Anchorage		33.8	39.1	45.0	10.9	5.0	1.81
Water transportation service		10.0	26.3	34.7	22.1	16.8	2.31

* Average ranking a (1) excellent to (4) poor scale. The higher the scale the lower the rating.
a Hispanic ranking < Anglo $t \leq 0.05$
b Hispanic ranking > Anglo $t \leq 0.05$
c Winter < Summer t = 0.05
d Summer > Winter t = 0.05.

Table 5. Part III Adequacy of services, facilities and programs.

1. In planning a visit to the Biscayne National Park, visitors usually expect that certain services, facilities, and programs should be available. Below is a list of statements about the Biscayne National Park. By checking the appropriate box, please indicate whether you strongly disagree, disagree, neither agree nor disagree, agree or strongly agree with each of these statements.

Statements The Biscayne National Park is...	EVALUATION (*)%					Average* ranking
	Strongly disagree	Disagree	Neither agree nor disagree	agree	Strongly agree	
A. A safe place to swim	2.3	8.0	15.1	57.1	17.6	0.94a
B. A clean, litter-free environment on shore	3.3	19.4	15.4	47.4	14.5	0.60
C. An enjoyable diving experience	0.5	3.2	19.4	52.8	24.2	1.20
D. A park with sufficient docking on the Keys	5.4	23.1	22.0	37.3	12.1	0.36
E. A park opening up nature's beauty	0.8	1.0	9.4	59.2	29.5	1.28
F. A park with convenient restrooms	5.0	18.8	25.8	40.4	10.0	0.42
G. A park offering enough direction signs	3.8	15.7	27.4	43.7	9.3	0.54
H. A park providing enough places to camp	4.1	14.9	40.4	34.3	6.3	0.40
I. A park offering a safe place for boating	1.6	5.0	8.7	61.9	22.8	1.09
J. A park adequately restricting commercial development on the Keys	4.6	6.6	14.3	42.6	31.9	1.06

Table 5 (continued)

Statements The Biscayne National Park is...	EVALUATION (*)%					
	Strongly disagree	Disagree	Neither agree nor disagree	agree	Strongly agree	Average* ranking
K. A park providing access to hiking, boating, fishing and active outdoor use	0.9	2.1	8.7	58.6	29.7	1.25
L. A park with reasonably priced services	3.8	7.6	31.4	45.5	11.6	0.78
M. A park providing a quiet, visual experience	0.9	2.8	9.6	59.1	27.5	1.21
N. A park offering an uncrowded boating experience	7.6	27.1	17.4	35.0	12.9	0.22
O. A park offering enough navigational aids, markers and flags	4.6	17.2	10.7	52.6	15.0	0.63
P. A park providing sufficient anchorage, docks and mooring buoys for boating, snorkeling and diving	3.6	23.3	15.3	45.4	12.5	0.47
Q. A park providing enough information on the weather, water conditions, life on the coral reefs, and what there is to do in the park	4.7	18.4	32.0	36.7	8.3	0.36
R. A park with sufficient shower facilities	6.3	20.9	42.7	26.2	3.9	0.01
S. A park with a welcoming atmosphere	2.7	4.4	23.8	54.5	14.7	0.97

* Weighted average with (–2) strongly disagree, disagree (–1), neither (0) agree (+1), strongly agree (+2). Positive scores denote agreement, negative scores denote disagreement.
[a] Hispanic > NonHispanic.

Table 6. Part IV Problems encountered by park visitors.

This set of questions concerns problems you and those with you may have run into during your last visit to Biscayne National Park. Please indicate how much of a problem each item was *for you or anyone with you on this visit* by placing a check ( ) in the appropriate box.

Problems		Not a problem at all	A small problem	A big problem
A.	Poor fishing	40.9	36.1	22.9
B.	Crowding on the Bay	42.4	42.8	14.8
C.	Litter on the Bay	37.8	43.9	18.2
D.	Crowding on the Reefs	59.9	30.2	9.9

Table 6 (continued)

Problems		Not a problem at all	A small problem	A big problem
E.	Pollution on the Reefs	54.3	29.1	16.5
F.	Crowding in campgrounds	63.4	26.5	8.1
G.	Noisy children in campgrounds	74.9	9.1	6.0
H.	Loud music in campgrounds	63.3	24.5	12.2
I.	Dirty restrooms and showers	47.1	33.2	19.6
J.	Not enough information about area's history	55.3	35.6	9.0
K.	Uninteresting presentation of the area's history	65.1	28.9	6.1
L.	Not enough information on park rules and regulations	56.5	29.9	13.6
M.	Not enough information on the area's natural environment	55.7	34.1	10.2
N.	Uninteresting presentation of information on the natural environment	68.6	26.2	5.2
O.	Not enough wildlife	57.4	29.8	12.8
P.	Too many insects	23.4	34.6	42.0
Q.	Not enough things for children to do	64.0	27.6	8.4
R.	Rowdy people	58.8	31.6	9.5
S.	Vandalism	65.2	25.9	8.9
T.	Pets not on leashes	73.9	19.4	6.6
U.	Not enough parking space near entry points	51.5	31.5	17.1
V.	Too many park service regulations on visitors	75.2	18.4	6.5
W.	Not enough written information about things to do and see	56.8	32.7	10.5
X.	Raccoons bothering campsite	77.2	18.9	3.9
Y.	Crowding at docks or in harbor	32.2	42.2	37.1
Z.	Reckless boat operators	24.2	38.5	37.1
AA.	Pollution in the Bay	36.6	41.6	21.8
BB.	Other (write in): 1. 2.			

Table 7. Part V Support for or opposition to proposed park changes.

1. At different points within the Park there are resources, facilities, and programs provided for you and other visitors. Sometimes visitors tell us that we need more or less of different resources, facilities or programs. Below is a list of things that people have told us we need more or less of – what do you think? By checking the appropriate box, please indicate whether you strongly oppose, oppose, neither oppose nor support, support, or strongly support each suggested change listed in the left-hand column.

		Degree of support or opposition( )				
Suggested change		Strongly oppose	Oppose	Neither oppose nor support	Support	Strongly support
A.	Need more places to get fuel and food on shore	12.2	12.6	24.2	32.6	18.4
B.	Need more self-guided tours on the Reefs	6.4	11.1	38.2	34.5	9.9

Table 7 (continued)

		Degree of support or opposition( )				
Suggested change		Strongly oppose	Oppose	Neither oppose nor support	Support	Strongly support
C.	Need more short hiking trails on the Keys	4.1	7.7	49.5	32.1	6.7
D.	Need more signs indicating where services are located	5.5	9.3	36.6	37.6	9.0
E.	Need more visible rangers patrolling	3.9	9.7	33.5	36.0	16.8
F.	Need more developed campsites	8.2	15.8	43.2	24.4	8.4
G.	Need more anchorage/mooring areas	5.0	9.5	33.1	38.6	13.8
H.	Need more rangers to inform us	3.7	10.5	49.4	26.8	9.5
I.	Need more park guides to explain things	4.1	11.6	54.9	23.0	6.3
J.	Need more picnic areas with toilets, water, shade, and close parking on shore	5.0	12.8	29.9	36.9	15.4
K.	Need more aggressive enforcement of safety rules and regulations on the Bay	3.6	9.1	31.4	31.9	24.1
L.	Need more clean restrooms with adequate supplies	2.9	4.8	30.8	41.0	20.8
M.	Need more navigational aids and markers	2.8	8.0	35.1	36.6	17.4
N.	Need more exhibits	4.6	11.6	55.2	23.5	5.0
O.	Need more live demonstration projects that inform us about the Bay	3.9	8.7	47.9	30.6	8.9
P.	Need more regulation of spear fishing	9.5	11.8	32.2	23.4	23.1
Q.	Need more visitor centers	7.3	15.5	52.8	19.7	4.6
R.	Need more regulation of boats around diving areas	3.5	7.7	24.9	37.2	26.7
S.	Need more restrictions on hook and line fishing	24.3	26.5	31.8	11.4	6.0
T.	Need more special use areas for snorkeling and scuba diving	4.4	8.9	27.4	37.5	21.9

Table 7 (continued)

		Degree of support or opposition( )				
Suggested change		Strongly oppose	Oppose	Neither oppose nor support	Support	Strongly support
U.	Need more areas set aside for just viewing the wildlife and their habitat	3.3	5.9	34.1	37.9	18.8
V.	Need more mosquito control in camping and picnic areas	5.8	6.3	18.0	27.0	42.8
W.	Need to charge fees to increase quality of programs	20.3	19.5	30.8	22.1	7.3
X.	Need less regulations and law enforcement on the Bay	29.2	28.8	26.7	7.8	5.7
Y.	Need less noise in campgrounds and along trails	3.3	5.9	65.9	18.4	6.4
Z.	Need less care in picking up litter	43.9	32.7	17.5	3.4	2.4
AA.	Need less regulations on diving	22.6	29.2	32.5	9.5	6.2
BB.	Need less water skiing	19.2	22.4	40.0	9.1	9.3
CC.	Need less sport fishing	39.1	28.8	24.3	4.3	3.5
DD.	Need less commercial fishing	6.8	7.2	16.5	17.3	52.1

Table 8. Part VI Interests.

We would like to ask you a few questions about the kinds of activities in the Biscayne National Park that interest you. Below is a list of activities that different people do. First, please check ( ) those activities that you or anyone with you participated in on your *last visit* to the Park. Use the box to the left of each activity for this purpose. *If you have not visited the Park, leave this column blank.* Then, *whether you have visited the park or not*, please indicate how interested you are (very interested, interested, or not interested) in each of the activities listed by checking the appropriate box to the right of each activity.

Participated on last trip (%)	Activity	Degree of interest % (Check the appropriate box)		
		Very interested	Interested	Not interested
56.9	A. Swimming or wading	62.0	32.2	5.7
42.0	B. Snorkeling on reefs	68.1	21.9	10.1
27.6	C. Diving	57.3	22.9	19.8
13.8	D. Spear fishing	31.1	20.7	48.3
46.3	E. Spin cast fishing	62.4	21.1	16.5
39.0	F. Visiting Elliot Key	44.9	45.9	9.1
11.6	G. Hiking	18.1	42.3	39.6
4.6	H. Camping on land	18.9	39.0	42.1
25.5	I. Picnicking	34.5	47.3	18.2
2.5	J. Traveling by tour boat	10.5	19.0	70.5
13.7	K. Sailing	33.4	25.3	41.3
61.9	L. Boating	78.2	18.8	3.0
16.3	M. Water skiing	33.6	29.2	37.2
12.8	N. Photography (land)	27.3	48.9	28.8
7.2	O. Underwater photography	26.3	36.5	37.3
22.8	P. Bird and wildlife watching	32.7	43.5	23.8
12.3	Q. Looking at visitor center exhibits	13.5	52.1	34.5
14.4	R. Nature walks (self-guided)	47.1	33.6	19.3
16.8	S. Boat camping	48.6	36.0	15.4
28.2	T. Using the park for access to the Fla. Straits for fishing	54.9	30.3	14.9
16.8	U. Beach combing	31.2	46.4	22.4
1.8	V. Ranger evening campfire programs	11.4	35.1	53.5
1.1	W. Ranger guided nature walks	10.7	34.2	55.1
1.3	X. Ranger guided snorkeling	14.6	30.1	55.3

Table 9. Part VII Outdoor activities.

Now we would appreciate some more information about the outdoor activities in which you participate. Below is a list of outdoor activities that different people enjoy. Please place a check in the box to the left of each activity that you participate in at least once a year. If you don't participate in an activity at least once a year, leave the box blank. Next, for each of the activities that you have checked (those that you participate in at least once a year), please indicate how often you participate in this activity by checking the appropriate box to the right of the activity.

Percent participate at least once a year (%)	Activity	HOW OFTEN? (%)					
		% Once a week or more	% 2 or 3 times a month	% Once a month	% Once every couple of months	% 2 or 3 times a year	% Once a year
32.4	A. Camping in remote wilderness areas	0.6	4.8	5.1	17.1	41.0	31.4
30.3	B. Camping in developed campground	1.3	2.0	6.9	14.7	47.1	28.1
15.1	C. Hunting	6.7	14.7	5.3	10.7	38.0	24.7
82.5	D. Fishing	20.4	37.9	17.1	13.8	8.8	2.0
14.7	E. Riding motorcycles, trailmobiles, snowmobiles, etc. off the road	8.8	17.0	6.8	23.8	24.5	19.0
11.0	F. Driving 4-wheel drive vehicles off the road	8.0	16.1	17.9	20.5	25.5	15.0
22.0	G Wildlife and bird photography	5.3	10.5	15.8	28.1	28.5	11.8
25.0	H. Bird watching	16.5	12.5	18.4	20.4	23.1	9.0
28.8	I. Hiking	4.4	7.7	16.8	27.5	33.6	10.1
37.8	J. Nature walks	3.4	8.9	16.3	26.8	31.8	12.6
42.0	K. Walking for pleasure	32.8	23.0	13.6	16.2	12.4	2.1
38.7	L. Bicycling	28.8	26.8	14.4	15.2	10.1	4.8
13.0	M. Horseback riding	7.1	7.1	10.2	9.4	30.7	35.4
21.4	N. Canoeing	2.3	3.7	7.9	23.1	32.4	30.6
31.7	O. Sailing	20.1	20.4	10.8	18.9	20.4	9.3
50.8	P. Other boating (water skiing)	18.0	32.3	17.6	16.6	12.6	2.9

Table 9 (continued)

Percent participate at least once a year (%)	Activity	HOW OFTEN? (%)					
		% Once a week or more	% 2 or 3 times a month	% Once a month	Once every couple of months	% 2 or 3 times a year	% Once a year
44.8	Q. Outdoor pool swimming	32.8	27.7	15.9	14.8	7.9	0.9
54.7	R. Other swimming outdoors	24.6	37.4	16.4	14.2	06.6	0.5
15.5	S. Golf	17.1	19.6	17.1	20.3	17.1	8.9
18.2	T. Tennis	24.2	20.4	14.0	19.4	15.6	6.5
31.1	U. Playing other outdoor games or sports	28.8	25.3	19.3	17.4	8.2	0.9
29.4	V. Going to outdoor concerts, plays	2.3	7.0	12.8	29.5	32.6	15.8
35.2	W. Going to outdoor sports events	6.8	11.8	15.1	30.4	29.0	6.8
58.3	X. Visiting zoos, fairs, amusement parks	1.8	4.4	10.3	24.5	40.5	18.7
45.6	Y. Sightseeing	7.1	16.2	15.8	25.8	31.2	3.9
47.7	Z. Picnicking	3.7	15.5	23.9	29.7	23.1	4.1
42.2	AA. Driving for pleasure	19.4	24.1	25.5	15.0	13.3	2.6
	BB. Other outdoor activities (write in): 1. 2. 3.						

FORMULATING POLICIES USING VISITOR PERCEPTIONS 251

Table 10. Section VIII Environment use and regulation.

In this section, we would like to get your opinion about a number of issues concerning the environment in general. For each of the following statements, please indicate the extent to which you agree or disagree by placing a check ( ) in the appropriate box

Statements	Strongly agree	Agree	Neither agree nor disagree	Disagree	Strongly disagree	Average*
a. We are approaching the limit of the number of people the earth can support.	18.6	26.7	25.1	24.8	4.8	0.39
b. The balance of nature is very delicate and easily upset.	42.8	45.7	6.2	4.9	0.4	1.35[a]
c. Humans have the right to modify the natural environment to suit their needs.	2.5	17.0	16.8	39.3	24.5	-0.80[a]
d. Mankind was created to rule over the rest of nature.	4.9	12.9	19.6	34.4	28.2	-0.85[a]
e. When humans interfere with nature it often produces disastrous consequences.	34.6	41.8	11.3	8.5	3.7	1.07[a]
f. Plants and animals exist primarily to be used by humans.	3.1	9.6	16.2	44.3	26.8	-0.98[a]
g. To maintain a healthy economy we will have to develop a 'steady-state' economy where industrial growth is controlled.	24.3	42.9	19.8	9.2	3.8	0.93
h. Humans must live in harmony with nature in order to survive.	56.7	36.9	4.4	0.9	1.1	1.54
i. The earth is like a spaceship with only limited room and resources.	35.5	40.7	12.6	8.8	2.4	1.12[a]
j. Humans need not adapt to the natural environment because they can remake it to suit their needs.	2.1	6.3	12.0	38.2	41.3	-1.25[a]
k. There are limits to growth beyond which our industrialized society cannot expand.	24.3	40.5	19.4	12.0	3.5	0.87[a]
l. Mankind is severely abusing the environment.	45.5	36.8	9.7	6.3	1.7	1.30
Summated scale	9.49					

* Weighted average with strongly agree (+2), agree (+1), neither (0), disagree (-1) and strong disagree (-2). Positive scores denote agreement, negative scores denote disagreement.
[a] Average for Hispanics ≤ .05.
[b] Average for Hispanics > Anglos; $t \leq .05$.
[c] Summated environmental awareness scale items A through L (C, D, F, H, J reversed). Range -20 (antienvironment) to +20 (pro environment). Hispanic 6.5, Anglo 10.0 ($t \leq .0002$).

Table 11. Part IX Conclusion.

These questions deal with information about you and your family. They will help us to determine who uses and who does not use the Park so that we can plan to meet your needs. *All* information that you give us is strictly confidential and will not be associated with you as an individual.

1. Where do you live? (Please fill in:)

    City:_____ State: _____Zip Code: _____
    (If outside USA: City: _____Country: _____)

2. Check the category that *best* describes where you have lived most of the time up to now:

Check one	Type of place
1.8	On a farm or ranch
3.4	In the country, but not on a farm or ranch
2.3	In a small town (population less than 2000)
6.8	In a town or small city (2501–25 000)
10.3	In a city (25 001–100 000)
37.5	In a suburb within 25 miles of a large city (more than 100 000)
38.0	In a large city (more than 100 000)

3. Is this place (where you have lived most of the time up to now) near the water? (Please check:)

    87.0 YES   13.0 NO

4. Check the category which includes your *total household income before taxes* last year:

Check one	Income category
1.0	Less than $5000
2.1	$5000 up to $10 000
7.7	Over $10 000 up to $20 000
14.4	Over $20 000 up to $30 000
19.8	Over $30 000 up to $40 000
17.9	Over $40 000 up to $50 000
16.9	Over $50 000 up to $70 000
8.7	Over $70 000 up to $100 000
11.5	Over $100 000

Table 12. Demographic characteristics.

5. What is the highest level of education you have completed so far? (Please check the most appropriate description.)

Check one	Level of education completed
1.0	Grade or elementary school
4.1	Some high school
8.9	High school diploma
15.5	Some school after high school, such as Business or Technical School
26.5	Some college
19.9	Graduated from college
9.3	Some graduate work beyond college
8.0	Master's degree
6.7	Doctoral or professional degree

6. What is your occupation or line of work? Describe what you actually do. (If student, housework, or retired, please say so.)

7. What is your marital status (Please check one:)
19.0	Single
69.2	Married
9.2	Divorced
1.4	Separated
1.3	Widowed

8. What is your race or ethnic background (Please check one:)

   Hispanic
   White, other than Hispanic
   Black, other than Hispanic
   Oriental
   American Indian
   Other (Specify: _____)

9. What is your age? (Please fill in your age:)
   Age: _____

10. What is your sex? (Please check:)
88.6	Male
11.4	Female

Conclusions

The case study of Biscayne presented above demonstrates a technique for evaluating visitor assessments of a marine recreational environment and how such information can be used to enhance both resource management and public policy guiding the evolution of the park system. From a park management perspective Biscayne's data suggest a need for sensitivity to expectations that different ethnic groups bring to the Park when designing services and programs offered within the Park. From a regional policy perspective, the data suggest addressing the issue of whether marine recreational areas should have increased development and formal control to maximize visitor satisfaction, or remain undeveloped natural areas.

Acknowledgement

The authors gratefully acknowledge the partial support of the National Park Service and the U.S. Department of the Interior in the conduct of this research. However, the opinions and conclusions are the authors' and do not necessarily represent official policy of the National Park Service or U.S. Department of the Interior.

References

Bauer, R. A. and Gergen, K. J. (1968). *The Study of Policy Formation.* New York: The Free Press.
House, E. R. (1980). *Evaluating with Validity.* Beverly Hills, California: Sage.
Latane, H. A., Mechanic, D., Strauss, G. and Strother, G. B. (1963). *The Social Science of Organizations.* Englewood Cliffs, NJ: Prentice-Hall.
Rossi, P. H., Freeman, H. E. and Wright, S. (1979). *Evaluation: A Systematic Approach.* Beverly Hills, CA: Sage.
Rutman, L., (ed.) (1977). *Evaluation Research Methods: A Basic Guide.* Beverly Hills CA: Sage.

Stephen V. Cofer-Shabica
University of Georgia,
Athens, U.S.A.

Robert E. Snow
Georgia State University,
Atlanta, U.S.A.,

Francis P. Noe
National Park Service,
U.S.A.

SECTION IV

Miscellaneous

22. Marine recreation in North America

Marine recreation in North America has undergone a very significant growth in the number of persons participating; but perhaps more importantly, in the diversity of activities in which they engage. This paper is in two parts. The first discusses the underlying cultural changes which support recreation. The second part analyzes the trends which have stimulated recreational pursuits and the growing number of industries which have characterized marine recreational activities during the past 15–20 years, especially as they have manifested themselves in the Eastern Seaboard on the U.S.

1. Historical and cultural antecedents

Leisure time and leisure time pursuits in North America have undergone a very subtle but significant change during the post WWII period. One of the main arguments presented here is that leisure and recreation, at least in an organized form, are relatively recent phenomenae in the U.S. for a very significant portion of the population. Because of the stronger European influence, these trends commenced earlier in Canada than in the U.S.

In order to trace these developments, it is necessary to go back in time to the early settlers, however briefly. Although the earliest East Coast settlements during the 15th and 16th centuries were numerically insignificant and were soon surpassed by subsequent immigration waves, it is suggested that the cultural impacts of the Puritans, Huguenots, Swedes and others were very significant and which can be traced to the present (White, 1967). These settlers not only succeeded in transferring the Judaen Christian work ethic to the North American shores, they nurtured it – perhaps out of necessity, considering the harshness of the new environment and the need to produce and store sufficient food and fiber to sustain the society during the winters and occasional disasters.

The need to husband or shepherd the 'natural' resources was seen by the early settlers as a mandate to control the environment. The classical literature exemplified by much of Hawthorne's work (1879) makes repeated reference to the foreboding forest and the need to suppress, control and manage this environment which, for most of the early settlers, was perceived as alien. Powell (1963), in a now classical study, reconstructed the early settlement of Sudbury, Massachusetts. This work verified many of the commonly held notions related to the early settlement patterns in the Northeast and the need to control the environment for reasons of initial survival, but perhaps more importantly, because of the perceived Christian edict requiring the believers to control and manage the hostile environment.

It is suggested that this need eventually turned into a demand for controlling and managing the environmental resources. Furthermore, this ethic was carried with the early settlers when the rest of the interior was settled. There is little doubt that the relatively rapid development of the country's natural resources sustained the economic development during the latter part of the 18th century and continued unabated through the 19th to the present. These developments were supported by technical development much of it underwritten by industry which invested capital in the creation of new technologies to control, (i.e., manage), the resource conversion process. While this thesis appears to have shaped much of the economic and technical development in North America, and especially in the U.S., it is cited here only as supportive evidence in explaining why recreation as a social right came late to the North American continent and only slowly became socially acceptable. The rapid expansion of the American economy during the 19th and first part of the 20th centuries was made possible by three sets of factors. The country's rich natural resource base was one factor. However, a rich resource base in and by itself is not sufficient to sustain let alone initiate economic and social development. There are still many countries whose resource base is better endowed than North America's, but have a standard of living substantially below that of the U.S. The second factor was of a relatively scarce labor force which was used intensively even during the industrial revolution. This argument is supported by the development of a long series of labor-saving devices in the U.S., the most important of which probably is the assembly line. Finally, there was a pervasive willingness and awareness of a need by both the private and public sectors to develop new technologies to overcome problems encountered in the economic development of the country. We have, in fact become a society which for much of its history has attempted to solve social problems through technical fixes.

By the early 20th century, North America was the sleeping industrial giant whose industrial potential had not yet been fully tested but which, within a twenty-year period, was called upon to support two World Wars with both personnel and material.

These accomplishments were made possible by a work force that, during the height of the industrial period relied on child labor and sweat shops where the laborers (whether man, woman, or child) were required to work long hours, six days a week, without the benefit of annual vacations – not to mention pension to sustain the worker during the all-too-often brief retirement period. It is suggested that the demand for vacations, and pensions began to gain acceptance during the second quarter of the 20th century. This may have been influenced by the social forces which impacted much of Europe during this period. The historic ties between Europe and North America, the mutual economic and military dependence, and the increased personal contacts enhanced by faster, cheaper traffic and communication between the two continents, no doubt played an important role in this transformation. The diffusion of the leisure time concept may have begun before World War I. The American expatriate writers and artists who included Fitzgerald, Hemingway, Gertrude Stein, and others contributed to the growing interest which many North Americans developed for anything European. It continued to grow between the two World Wars as trade, initiated especially by

American companies, as they expanded through the creation of subsidiaries on the European continent. Many of the millions of American servicemen who spent time in Europe during and following both wars no doubt became aware of the mandatory vacations which most European laborers were entitled to even then. Since the end of the hostilities, the notion of paid vacation and travel for pleasure grew in North America at an almost unprecedented pace.

The manner in which this leisure time has been expressed differs quite significantly from the way in which Europeans spent and utilized leisure time. Several factors contributed to the recreational patterns which evolved in the U.S. These include the vast expanse of the U.S. (much of it very sparsely populated), greater mobility by a comparatively larger proportion of its population, an early initiative to set aside extensive areas of unique environments, and a greater number of the population owning or occupying single-family homes.

Leisure patterns have changed dramatically since the end of the Second World War, which have had major impacts on the outdoor recreational facilities located in the interior of the country and along the coasts. These changes have been influenced by the manner in which leisure time has become available. Each of these is discussed below.

As the obligatory workday was reduced from 10–12 hours to between seven and eight hours a day, it increased the demand for open space in close proximity to urban areas where fewer people were living in single-family houses. The usage rates in city parks increased along with the demand for additional open space in close proximity to the home and work place. These demands resulted in the development of urban pocket parks and in a few instances waterfront developments including, recreational piers (Clawson, 1963). In suburban and rural areas, enormous efforts and vast amounts of money were spent on beautifying private gardens, probably the most important outdoor recreational outlet available to a large segment of the American public.

The five-day week is now almost universally accepted in North America. This development has put pressure on intermediate facilities located some distance from major urban centers. Clawson is of the opinion that many of these facilities are state and regional parks located within 100 miles of the user's residence. In recognition that most of the areas set aside by the Federal government were located in the more sparsely populated West, (and de-facto inaccessible to many less well-to-do urban residents), Congress established several National Seashores, and National Recreational Areas most of which were built in close proximity to urban areas. The phenomenal growth in water-related activities is a reflection of this demand, a topic which will be returned to subsequently.

The third area impacting outdoor recreational areas in general, and marine recreational facilities in particular, is the increase in annual paid vacations by a growing number of Americans. Comparatively speaking, this development came relatively late to the U.S. Even today, unlike much of Western Europe where annual paid vacations are at least one month, U.S. employees are generally limited to two weeks.

The impacts of annual vacations has been felt both domestically and internationally. In the U.S., the development of vacation homes, condominiums, and timeshare facilities which started in the South has diffused to the more temperate shorelines. Another development is the increase in trailer parks, coastal hotels, motels and, most recently, the cruising vacations on cruise ships and on smaller pleasure crafts. International tourism has increased in both directions for some time now. While significant perturbations have taken place as a result of currency exchange rates and security related to international terrorism, leisure travel is likely to increase both domestically and internationally for the foreseeable future.

Two partially independent factors account for the fourth and final increase in leisure time. The U.S. population is growing older. While the median age is increasing, in large part this is due to smaller and delayed families and to rapidly declining birthrates. The full impact of these developments is not likely to be felt for some time to come, even though some manifestations of these developments are already clearly in evidence. Of greater direct impact is the increased life expectancy for both sexes and the growing number of Americans who now have additional financial support above and beyond that offered by Social Security. An additional factor is the increasing number of persons who are opting for early retirement, enabling them to participate in fairly active recreational pursuits often for several decades following retirement.

The increase in leisure time in all four modes and the corresponding improvement in economic ability and overall health have compounded the impact on the neighborhood intermediate- and resource-based recreational facilities. However, the more important manifestation of the retired population has been the migration of many retirees to the sunshine and adjacent states. For some, this move has been permanent, restricted to so-called retirement communities. For others, especially the younger generation, this has been a partial movement where two residences have been maintained – an annual migration south in the Fall followed by a return to the north during the Spring.

The rapid population growth in Southern California, Florida, Texas, Arizona, and New Mexico which took place during the 1960s, in large part was affected by this migration. For a variety of reasons, a switch from the traditional sun states appears to have taken place during the 1970s and early 1980s where the Mid-Atlantic states (Georgia, North and South Carolina, and Virginia) experienced the most rapid growth (West, 1978).

There is little doubt that the cultural changes that have taken place in the U.S. especially since the end of the Second World War have contributed in a major way to the rapid expansion and pressure on coastal resources; yet these demands were primarily spatial in the sense that usage was felt on both existing and new facilities. Clearly, the demand for increased public access is a case in point. Public access is but one of several topics which were highlighted and subsequently de-emphasized in the coastal zone management plans that now cover over 80% of the U.S. coastline. However, equally important is the demand for equipment and facilities which the private sector has successfully sought to fill.

2. Marine recreational developments

What specific impacts have the historical antecedents had on marine recreation in North America? Several factors bear on this issue, including the development of new materials and technologies and the fact that many water recreational activities were developed by the user. The first two sections will address these factors which will be followed by an analysis of the demand for marine related recreational activities and the supply of equipment and facilities.

2.1. The impact of technology

The development of space-age technologies and materials has had two major impacts on outdoor recreation, especially marine recreation. New materials have enabled the development of equipment which, just a few years ago, was unavailable to the average user; and new activities made possible the creation of new equipment.

The development of new materials has been especially noteworthy within the boating industry where weight and strength are of prime importance. More recreational boats are venturing offshore than at any time before. Moreover, because of the inherent strength in some of these materials, the vessels which are built today can accommodate more people in relative comfort and in spaces which are significantly smaller than a generation ago. Kevlar, mylar, carbon reinforced fiberglass and titanium, while still expensive, are becoming more available as building materials in the boating industry. It is reasonable to expect that many of these new materials will be as common in the future as wood, aluminium, and fiberglass are today.

Not only has the boating industry been affected but other marine recreational activities as well. Forty years ago, scuba diving (Self Contained Underwater Breathing Apparatus, first developed by Jacques Cousteau), was viewed as a scientific tool not generally available to the public. The Second World War and the highly publicized exploits of the 'frogmen' teams glorified in films, and later on in several very successful TV series (some sponsored by the National Geographic Society) undoubtedly did much to help popularize recreational diving, which claims between four and five million participants in the U.S.[1]

The second development likely to result from the introduction of new space-age materials is the widened rate of participation. New materials, such as fiberglass, have made it possible to develop equipment (i.e., activities) which were impossible to mass produce in the past. Windsurfing, surf-flying, parasailing, are a few of the activities which owe their existence to the development of new materials.

Finally, a word about the future. Outdoor recreation in the U.S. has traditionally enabled participants to develop new activities dependent upon interest, availability of location, etc., with little or no interference on the part of the government. Only in those instances where danger is apparent has the government become involved, preferring to let the industry regulate itself. The aforementioned diving certificate is a case in point. At this time a rather lively debate is taking place which addresses the issue of licencing maritime recreational fishing, operators of off-road vehicles (ORV) and powerboats.

From a marine recreational planning perspective, allowances must be made to permit future uses, including recreational, of the coastal and nearshore marine environment.

2.2. The evolution of marine recreational activities

With few exceptions, the marine recreational system is user stimulated, which means that innovations, techniques and uses originate with the user – unlike activities promoted by the equipment or facility provider. While there are exceptions, especially within the more expensive activities like boating and diving, the majority of new coastal and marine recreational pursuits originated among innovative users. The Hawaiian surfboard is a case in point. It is not surprising that surfing originated in Hawaii given its ideal conditions for this activity. The idea of surfing soon diffused to other areas on the mainland and overseas where breakers of suitable heights were present. Because of the nature of the sport, traditional surfing could not be undertaken on low energy beaches. Nonetheless, it did not take long to transfer the surfing concept to areas where wave height was insufficient to enable participation. In low-wave-energy environments (lakes, rivers, and protected bays) the motive force became the relatively small but powerful outboard-driven powerboat.

Shortly after the Second World War, it became fashionable in certain areas to pull a board behind a powerboat on which the 'surfer' would ride. The surfboard, in turn, evolved in two quite separate activities: waterskiing and sailboarding. Waterskiing traces its origin to the surfboard pulled behind the powerboat. Waterskiis have the advantage of enabling the participants more freedom of movement compared to the boat pulled surfboard. The development and implementation of new uses of the marine environment did not stop there. Parasailing is another activity which counts a small but devoted following and which combines elements of hangliding, parashuting, waterskiing and powerboating in one activity.

The evolution of the surfboard into the sailboard is equally innovative and represents a case where the surfboard could be used in environments without waves. The driving force of the board was changed from wave energy to wind energy. Coincidentally, the development of the sailboard also represents one of the few instances where a water-based marine recreational activity has been transferred to dry land. The modern skateboard, currently very popular among the young, is directly related to the traditional Hawaiian surfboard. Two additional developments may round out the evolution of these activities. In a few areas, skateboards have added sails in an attempt to let the wind drive the board, and a few sailboard enthusiasts have replaced the wheels of the skateboard with ice skates and returned this concept to the frozen marine and lake environment. In fact, this development may be the true skateboard as it perhaps was originally intended by the inventor.

While few of the more extreme developments have attracted a great following, they are cited here as examples of how the supply of marine recreational facilities and equipment has been driven by consumer innovations. The next two sections analyze a) the demand for marine recreational activities, facilities and equipment, and b) the supply of these.

2.3. Demand for marine recreation

Traditionally, marine recreation has been synonymous with fishing, beaching and bathing. While still very popular and growing, there have been some important developments in these areas. The demand for beach and bathing facilities has largely paralleled the demographic developments. It is now possible to identify two separate beach-demand systems – one related to urban areas, and one associated with suburban middle income households. Urban beaches are increasingly seen as the single-most important recreational outlet for a large segment of the urban population. This has placed severe socio-environmental demands on those beaches accessible to public transportation. Examples include the famous Coney Island in New York City, and Revere Beach in Boston. During the 1960s, prior to the implementation of Sect. 208 of the Clean Water Act, many urban beaches were threatened by high coliform bacterial counts making them a potential health hazard. While still of considerable concern to some urban health officials, it is quite clear that beaches with respect to organic pollution in U.S. coastal waters, appear to have stabilized. In a number of instances they may even have improved.[2] Many of the urban beach facilities are overcrowded and quite a few suffer from budgetary constraints, resulting in poor maintenance and problems associated with dilapidation of the physical infrastructure. During the 1970 decade, this problem was recognized by the Federal government, which proceeded to fund the development of several urban-oriented recreational facilities. The Gateway National Recreational Area in New York, (Foresta, 1985), and the Golden Gate National Recreational Area in San Francisco are two coastal examples (Scott, 1985). These developments were very severely curtailed during the 1980 decade when emphasis was placed on the development of existing land resources as opposed to protecting new areas under the aegis of the federal government.

The demand by the middle class has had both social and geographic implications. Beach facilities which are not accessible by public transportation are used by a) higher-income groups with their own modes of transportation, and b) by groups requiring special needs. Finally, a growing number of the middle-class population is seeking beach and bathing opportunities as tourists in tropical and subtropical environments away from the home during the winter season. Each of these will be discussed below.

There is considerable evidence that a functional segregation of beach facilities has taken place in recent years. This has been influenced more by accessibility[3] and less by access. Those with private cars (i.e., the middle class) have been able to travel to beaches which are especially suited to meet their specific needs and expectations.

This functional segregation of beaches is based not so much on gender and race but by the specific needs of the beach-using public (Heecock, 1966). While no scholarly paper has yet been written concerning the issues surrounding 'clothing optional' beaches, this phenomenon in North America still represents a difficult-to-deal-with problem for coastal resource managers in charge of managing state and Federal beaches.

On a more formal note and in response to P.L. 94–142 often referred to as the Civil Rights Act for the Handicapped, a growing number of facilities are now becoming accessible to the physically handicapped. Furthermore, this development is not restricted to beaches but to a varying extent is being required for all facilities involving Federal funds, permits or property.

The increased ability to travel for leisure outside of their respective countries is possible for many Canadians and Americans, some of whom visit subtropical island nations in the Caribbean and the Pacific. This development, which is concentrated during the winter has had a rather significant impact on both the domestic and international island nations and established a pattern which has been replicated initially by private boatowners living in the South and the Caribbean wishing to earn some money by leasing their boats for short or long periods of time. An industry has been created offering charter services in nearly every major tropical and subtropical ocean.

The influx of tourists in several of the smaller islands has not been without some deleterious side effects. Most popular tourist sites have a limited season; thus, the tourist-stimulated businesses are usually cyclic and often unable to sustain the local economy during the off-season (Matley, 1976). Of potentially greater significance are the impacts on both the physical environment and the municipal infrastructure, such as sewage treatment, demand for water, etc. (Mathieson and Wall, 1982). Since most tourist facilities require services comparable to those the tourists expect from their home environments, these services often place an unreasonable demand on the local resource base, to the point where it is unable to sustain the local economy. The demand for water to irrigate golf courses in some of the islands located within the tropical savannah climate (Aw) may deplete available water supplies which are needed to sustain local subsistence farms. These problems are not limited to the Caribbean and tropical islands in the Pacific Ocean, but have been identified in the Mid-Atlantic coastal states as well. This geographical region was during the latter part of the 1970s, the fastest growing (West, *op. cit.*). The high density development of Hilton Head, SC, has resulted in a number of socio-environmental impacts. The watertable underlying Hilton Head Island has become severely depressed to the point where it now extends beyond the Calibogue Sound and affects the coastal areas on the mainland. This lowering of the watertable is referred to as the depression cone (Gordon, 1983). Other impacts have increased the local taxes to the point where the original black population has been forced off the island.

2.3.1. Fishing. Recreational fishing remains one of the most popular marine recreational activities which has also undergone significant changes in recent years. Bryan (1976) estimated that 33 million participated in the fishery, 40% of whom fished from shore. Nearly one and one-half billion lbs. of fish were caught equalling in weight the catch by the commercial fishery. Each participant fished on the average of 20 days a year with each trip lasting between 4–5 hours. He also makes the observation that a disproportionate small number of fishermen caught most of the fish.

The discussion of this activity is divided into shore-and-boat-based. Of the two, there is little doubt that shore-based fishing is the most diverse and popular, attracting participants from all walks of life and involving equipment ranging from a line rolled around a beer can to expensive surfcasting rod and reels costing several hundred dollars.

During the 1985–86 sampling year, the National Marine Fisheries Service estimated that nearly 5.5 million persons fished along the Maine to Florida Key shorelines. While fishery statistics are notoriously unreliable, the NOAA estimate of participants is impressive. With few exceptions, this activity coexists with an aggressive commercial fishery, especially in the Northeast, with comparatively little conflict (Duell, 1977). This is due, in large part, to the nature of the commercial fishery which is based primarily on schooling fishes numbering about 100 species out of the approximately 1000 present on the Atlantic shoreline. Of these, nearly half (some 500 species) are sought after by recreational fishermen (Freeman, 1977).

Although firm growth estimates are virtually non-existent it is estimated that participation in recreational fishing is increasing about 10% yr. Two aspects of the shore-based fishing are of considerable geographic importance, although the tentative conclusions are based on case studies.

First, most of the population living along the Atlantic Seaboard is urban, and most are living in countries bordering the ocean (West, *op. cit.*). Consequently, it is to be expected that most of the shore-based fishing occurs within a relatively short distance of urban areas. It is clear that a considerable amount of fishing activity does take place along urban waterfronts, perhaps enhanced by a perception of improved water quality. While the average catch, a point which will be returned to below, appears to be quite low (Heatwole and West, 1984), several investigations have found that people participate in fishing for reasons other than catching fish. Spaulding (1976), in a study on the Narragansett Bay recreational fishery, found that only 34% of the respondents he interviewed fished for the catch. This category was superseded by those respondents who felt that relaxation associated with fishing was more important (38%).

Second, brief mention was made of the improved water quality in many urban areas as related to fishing. These improvements have been limited, however, to the organic biodegradable pollutants and not to the persistent (non-biodegradable) pollutants including heavy metals and compound organics like PCB, Eldrin, DDT, etc.

Physical scientists (Belton et al., 1986; Cabelli et al., 1983), and social scientists (Moser, 1987; Cable et al., 1987; Dinius, 1981; David, 1971), have addressed the potential risks associated with ingestion of fish and the awareness (perception) of nearshore pollutants. This problem may be accentuated among recent immigrants, especially from Southeast Asia and the Caribbean, many of whom used to fish in their native countries, a practice which has been continued in their adopted country. The potential problems related to this group of recreational/subsistence anglers concern the catch composition and disposition of the catch. Many Southeast Asian fishing in urban areas tend to keep a greater proportion of what is being caught. In other words very little if any of the catch is being thrown back including shellfish and other species which spend most if not their entire lifecycle in the nearshore much of which is heavily contaminated. Since some of these species are suspect from an epidemiological point of view, there is some concern that build-up of potentially dangerous pollutants may occur among this population (Heatwole and West, 1983).

2.3.2. Passive coastal recreational pursuits. These activities include those in which the participant uses the environmental resource without directly consuming and/or contaminating the resource. Examples include sightseeing, walking, jogging and other similar activities which could be carried out in non-coastal environments but which are enhanced by the shoreline environment. Consequently, the impacts associated with passive uses are likely to be fewer, although rarely totally absent.

The relatively fewer impacts associated with these activities compared to active coastal marine recreational pursuits, refer only to the activity itself and not the facility within which it is being carried out. Consequently, the scenic road which may be built in a coastal environment to enhance tourism and recreation could have resulted in fairly substantial adverse shore and nearshore environmental impacts. This point is particularly relevant if one includes the construction of coastal roads and secondary homes both of which have occurred in numerous coastal counties in North America and elsewhere.

Unfortunately, there is very little data available addressing passive recreational demands in the coastal region. This is particularly surprising, since such demands are often cited as one of the most important secondary reasons for constructing roads and other public work projects in the coastal zone.

Passive recreational pursuits have often been the *raison d'être* for undertaking urban waterfront renewal projects. These projects have borrowed heavily from the integrated commercial mall complex which has characterized shopping in suburban areas in North America. The prototype of a North American waterfront redevelopment scheme consists of refurbishing one or more older buildings – usually warehouses and/or the construction of new buildings which, from a design point of view, are intended to blend in with the existing structures.

The activities associated with such projects are not only people oriented, but tend to target the socially upward mobile population. Businesses tend to be small and situated in an environment with a marine ambiance which often has little to do with the working waterfront. The more successful projects are integrated developments which include, besides businesses, restaurants, upscale residential projects, international trade centers, maritime museums, acquaria and marinas. Some of these projects have been very successful commercially and have revitalized many surrounding areas. On a national scale, urban waterfront renewal efforts probably amount to no more than one percent of the total urban waterfront, however, the secondary impacts on adjacent neighborhoods have often been considerably greater than the governmentally sponsored waterfront project itself. To some extent these projects have been viewed as stimuli which economic impacts were to diffuse along the waterfront as well as inland.

2.3.3. Water activities. If data on shore-based recreational usage is sparse, then data on nearshore activities is virtually non-existent. This seems especially surprising considering the territorial jurisdiction of the Coastal Zone Management Act (CZMA, 72) which covers both the nearshore and dryshore. As stated above, the recreational use of the nearshore has grown in response to increased standards of living, more and diversified uses, and material technology developments.

Only a handful of papers have addressed the recreational uses of the nearshore. West (1982) developed a classification of potential conflicts among various users of the marine environment and attempted to address these in the context of the spatial jurisdiction of this environment. These jurisdictions include the Army Corps of Engineers (ACE) designated harbors, the U.S. designated inshore waters, the portion of the territorial sea extending from the baselines and/or outer boundary of the inshore area to the end of the territorial sea, and finally, the area where the international collision regulation (COLREG) prevails. Chaney (1979) sought to identify boat use in Maryland, while Oregon and a few other states are in the process of developing a very ambitious and far reaching effort of zoning the Territorial Sea.[4]

The resource use problems encountered in the nearshore are uniquely geographic, compounded by the fact that many uses involve a three-dimensional environment, unlike most land-based uses which occur within a cartesian plane. Furthermore, most marine recreational uses are dynamic in the sense that the use of the resource involves movement horizontally across the surface or bottom (diving), or vertically through the water column (boating, fishing and diving). A further complicating factor relates to the multi-use nature of most recreational boating activities. Few boaters are single-use consumers of the marine environment but engage in several different activities while on the water. Some of these are passive and related to the amenities associated with viewing the water, shore, and other boat-related activities; while others are active, involving only one use, e.g., waterspace for waterskiing, fishing, etc.

There is little doubt that fishing is the most popular boating pursuit. This can take place from small skiffs which are cartopped to the preferred fishing site to large power boats which cost in the hundreds of thousands of dollars. Sportsfishing from boats, whether privately owned, chartered, or taking place from a party (head) boat is affordable as an activity to nearly everybody. Half-day trips on party boats may coast as little as $20 for a four-hour trip. Depending upon species, it is not unusual for the successful angler to land a catch in excess of the cost of the trip.

As stated above there are many reasons why people engage in fishing as a recreational pursuit, and nearly all of these suggest that catching fish is of only secondary importance. Nonetheless, the probability of catching something appears to be an important factor influencing the individual's decision-making process. Thus, any effort made by the fish manager, charter, party or private boat captain to improve catch rates would enhance the fishing experience and the economic benefits to the businesses supporting this industry. Marine biologists have long been aware that reefs – whether located in the temperate, subtropical or tropical environments – tend to enhance the environmental conditions for plankton and higher forms of marine life, including fish. As a result, for 200 years, the Japanese have used the artificial reef concept to enhance commercial fishing.

Buchanan (1972) found catches over artificial reefs to be 10 times higher compared to the natural non-reef habitat, a finding which was verified by Sheehy (1986). There are many unanswered questions related to the construction and placement of artificial reefs and their overall impact on the total biomass. One such concern is their overall impact on the biomass. Do artificial reefs simply concentrate the fish in areas where they are easier to catch, or does the construction of new artificial reefs increase the total biomass? The definitive answer to these perplexing questions have yet to be written. Nonetheless artificial reefs are being developed in increasing numbers, especially along the south and mid-Atlantic coasts, with fewer created along the Northeast coast.

The primary purpose in developing these structures has been to enhance marine recreational fishing; in fact, the first artificial reef structures were developed by private sportsfishing clubs to enhance fishing. Since the passage of the CZMA in 1972, sites located within the coastal states' territorial sea jurisdiction are now subject to the rules and regulations written in support of the coastal states' coastal management program.

Artificial reefs fall into two major categories depending upon intended usage. The deeper reefs are constructed primarily to serve the boat fishing community, while the reefs located in the more shallow waters are intended primarily for SCUBA divers (Gardozo, 1985). Other reef classifications have been devised based primarily on construction material and type, since preliminary research suggests that different materials may affect colonization and thus the species likely to occupy the structure.

Other boating activities include a plethora of activities which only commonality is a shared environment where the activities are carried out on some sort of floating platform which is driven by either an internal or external force. The specific boat-related activities range from long-distance endurance races on board either powerboats or sailboats through a host of different boating activities where the primary activity is the movement of the vessel, to activities where the vessel plays a comparatively minor role in the activity. Specific boat-related activities include, besides those mentioned above, cruising nearshore and offshore, daysailing, and racing. The activities where the vessel serves as a platform for the preferred activity include, (besides fishing), hunting, diving and houseboating.

What common trends characterize all boating activities, whether primary or associated activities? For reasons outlined above, it is reasonable to expect that boating will continue to grow in popularity, although not all activities will experience comparable growth nor will all areas be equally impacted.

The following discussion is divided into two sections. The first concerns those activities where the movement through the water represents the primary purpose of the activity (hereafter referred to as 'primary boat activities'). The second includes those activities in which the boating platform serves as a place on which the primary activity takes place (hereafter referred to as 'boat-associated activities').

The growth in primary boating activities has been subject to considerable fluctuations between powerboats and sailboats and small versus large boats. While no studies have been undertaken identifying the reason(s) for slower growth in sailboat sales, it is clear that the majority of new entrants to the activity have opted for powerboating as opposed to sailboating. Several reasons may be suggested for this trend. Some of the more common include the higher level of knowledge to be able to utilize wind as the primary motive force. Another often cited reason relates to the inability of many to cope with the whims of weather and climate. Busy schedules require some degree of reliability concerning estimated time of arrival (ETA). Other reasons cited are the psychological need of many to make their own decisions; Using wind as the primary energy source makes it difficult to satisfy these.

A third argument setting power and sailboaters apart relates to how the benefits associated with the various types of boating are being perceived. It is possible that the sailboat operators derive most of their pleasure during the trip itself while the powerboaters may place a higher priority on the activities associated with the point of destination? While these are interesting research propositions, they are just that. No comprehensive research has been completed addressing these and other comparable questions.

Another trend which has been in evidence is the preference for the larger boat (both power and sail) by most boaters. In other words, that section of the boat building industry which concentrates on larger vessels continues to do relatively better compared to those emphasizing the smaller vessels.

Boat-associated activities exhibit a wide range of water recreational pursuits, including diving, waterskiing and a wide range of fishing modes. To some extent, the boat serves two primary purposes: as a means of transporting the participants to the preferred site, and as a platform on which the preferred activity can be engaged in. One should not infer, however, that the boating part is not enjoyed. For most boat-associated recreational activities the trip to the site is considered an integral part of the activity, albeit perhaps not the most important.

Of the boat-associated activities, fishing and diving (SCUBA) are the largest boat-associated marine recreational activities. Since fishing has already been discussed, the following emphasizes SCUBA diving, most of which takes place from a boat. As mentioned above, SCUBA divers are a large marine recreational group who continue to grow at an annual rate of about 240 000 (Matheusic and Mills, 1983).

Recreational diving is less than 30 years old. Even so, this activity is as diversified as boating itself. Many recreational divers engage in this activity passively in the sense that the object is to observe the marine environment. A somewhat smaller group combines fishing (spear) with diving, while other divers are collectors, looking for anything from biota and rocks to cultural marine artifacts.

The latter activity has lately become quite controversial since many divers are looking for and seeking artifacts of historical and archaeological value, thereby coming into conflict with both the marine archaeologists and commercial salvor. This problem is a classical coastal resource problem; the only difference is that the resource is located in the marine portion of the coastal zone, unlike most other coastal management problems which have addressed terrestrial issues.

Most of the more than 12 000 known shipwrecks in U.S. waters are located in relatively close proximity to the coastline (within the territorial sea, Gieseck, 1987), where many have met their demise because of reefs, congestion or similar calamities.

Bascom (1976) suggests that 40% of Eighteenth and Nineteenth century wooden ships ended their careers by grounding. This estimate apparently does not include sinkings due to other causes. Furthermore, wrecks appear to be concentrated in approaches to present and formerly important harbors. Some 300 known wrecks have been mapped in waters surrounding Block Island. This is a small glacial morrainic island seven miles long and three miles wide located at the entrance to Long Island Sound, Narragansett, and Buzzards Bays.

The territorial jurisdiction of shipwrecks adds another interesting – albeit complicating – dimension to the management of this resource. Wrecks located within the coastal states' territorial sea fall within that state's jurisdictional domain, provided the state has passed legislation covering the exploitation of marine archaeological resources. Exceptions to this rule include navy vessels, whether domestic or foreign and certain other civilian vessels which have not officially been declared abandoned. Of the 36 coastal jurisdictions in the U.S., twenty-seven states, territories, and commonwealths have enacted legislation to manage access and disposition of this resource (Giesecke, *op. cit.*).

The position of the recreational divers with respect to access and excavation is in all probability overstated, considering the size of the resource base and the relatively small proportions of valuable wrecks. Several leading marine archaeologists have involved recreational divers in many of the excavation projects undertaken in recent years. In part, this development has been caused by the absence of qualified marine archaeological divers and the interest which many divers (and other marine recreationists) have in participating in legitimate projects. Other groups which have contributed to the advancement and welfare of the activity they expound include the various boating auxiliaries which serve under the sponsorship of the U.S. Coastal Guard. The diving certificates which have been initiated by industry or interest groups include the following: Professional Association of Diving Instructors, (PADI), the National Association of Underwater Instructors, (NAUI), and the YMCA diving certificate are all examples of how individual organizations with an interest in the promotion of the activity have established standards intended to serve the welfare, health and safety of all participants.

Although this discussion has attempted to summarize the conditions prevailing at the present time, it may be fitting to conclude by reiterating one point which has already been made. Marine recreational geographers and planners need to take into consideration not only the demands of those participating in activities known to us today, but also be cognizant of the environmental requirements of new activities which no doubt will be developed in the future. This can only be accomplished by undertaking a continuing process of evaluation of current coastal recreational demands and match those to the available supply of facilities accessible to those demanding them. The next section discusses the supply side of the recreational system.

2.4. Marine recreational suppy system

The supply side of marine recreation can be broken down into two major groups. One of these refers to the vendors of the facilities used by the public, while the other group consists of the specific services, facilities and types of equipment used by the public.

Unlike the recreational demand discussion which addressed the issues related to each of the major marine recreation activities, this discussion is organized according to what is required in terms of services, facilities and equipment, and which sector is providing it.

In the past, nearly all equipment required to engage in marine recreation was acquired by the user. Thus boat, fishing poles, diving equipment, and more recently windsurfers, etc. were generally purchased or built by the participants. Because of the relatively short season – at least in the North East – and the high capital investment of some of the equipment, such as seaside condominiums, boats, etc., there has been a slow but subtle move towards renting and leasing (charter) some or all of the equipment required. This trend has lowered the cost on a per-use basis, which has enabled a greater number of people to participate.

Time sharing of oceanfront properties has enabled a much greater number of people to enjoy the amenities associated with the marine environment albeit not for the longer periods which traditional owners have had. The virtual explosion of commercial charter boats (bareboat or fully crewed and supplied) that have been established throughout the Caribbean, Mediterranean, and increasingly in the Pacific has broadened the demand for boating experiences in regions which just a few years ago was limited to a very select few.

The concept of timesharing and chartering has been further developed in Florida, where at least one marina operator is now leasing small sailboats on a walk-in basis – quite similar to procedures engaged in when renting a car. There is little doubt that the leasing/renting market of recreational equipment has been a major factor in the continuing growth of the marine recreational business. This industry commonly ranks first or second in most coastal states on the eastern seaboard.

The marine recreational equipment required to engage in the activity is almost exclusively provided for by the private sector. Since a commercial market can be established, there has been no need for the public sector to act as a vendor.

The question concerning which of the two sectors should provide the services which may be required by a given recreational group is not as well defined. In general, where the service is marketable, the private sector will tend to eventually fill the market, although in a few instances the initial provider may rest with the public sector. For many years, Search and Rescue (SAR) has been the exclusive responsibility of the United States Coast Guard (USCG), a responsibility which has largely been turned over to the private sector in those instances where no threat to the health and welfare of the boater is present. In the latter case, the USCG will still engage in SAR efforts.

Other services are still provided by the Federal government, such as navigational aid. Nonetheless, for several years now, efforts have been made to transfer a portion of the operating cost of providing buoyage to the using public (commercial shipping, fishing and recreational boating). If enacted, this would be implemented as an annual user charge or fee (tax) to be paid to the federal government for the privilege of using the marine environment. As currently envisioned, the USCG would still be in charge (and responsible) for maintaining the overall integrity of the navigational system, although components of that may be contracted out to private vendors.

Although the examples cited here refer to services provided on the Federal (national level), comparable services exist on many state and some local levels as well. Most states maintain boat registration systems, which original purpose was to insure against theft and serve as an identification and verification system in case of damage and loss. Because of the growth in both boat inventory and increased boat registration fees, many states now treat this fee as general income as opposed to designating the money for boat improvement projects. Marine recreational services offered on the local level include lifeguards on many local (and state) beaches, informational services directed to tourists, and day trippers to specific sites.

Some services are offered by the private sector as well, although most of these have been developed for purposes of stimulating interest in a given product or activity. In conclusion, there appears to be a growing tendency to commercialize services in those areas where the service is marketable, or where the jurisdictional powers have the ability to tax or extract dues for services used. To some extent, this represents a rather marked deviation from previous practices where most marine-related services offered on the Federal levels were provided free of charge.

The final supply category includes the facilities within which the recreational activity occurs. Because of the nature of the subject matter, these are located on land and include the entire nearshore marine environment, although in the latter case not generally as an exclusionary facility.

The jurisdictional issues concerning the nearshore area are dependent upon location and activity, as well as which part of the marine environment (surface water column or bottom) is being used. In general, the Federal government retains jurisdiction for the purpose of navigation in areas extending from the shoreline to the end of the Territorial Sea, commonly defined as extending three nautical miles seaward of the baseline. However, the coastal states retain jurisdiction over the resources located[5] on the bottom and within the water column, and therefore has wide latitudes in setting fishing standards within this area.

Some coastal states have taken steps to measure the impact of recreational fisheries on the resource base; however, most coastal states rely on the National Marine Fisheries Surveys which are published approximately every year. Several coastal states require licences for recreational anglers, although this practice is not very widespread on the Atlantic Seaboard.

One of the problems concerning coastal states' involvement in the management of the fish resource relates to the relatively few species which spend their entire lifecycle in inshore and nearshore waters. On the Atlantic coastlines, the only important fish pursued almost entirely by the sportsangler is the striped bass. The management problem associated with this resource has been one of coordinating the management schemes of adjacent coastal states through which the striped bass migrates. This problem has partly been eclipsed by the discovery of high PCB levels which make consumption of this fish problematic for the foreseeable future.

The management of biological resources seaward of the Territorial Sea to the outer boundary of the Extended Economic Zone rests with the Federal government. To manage these resources, a number of Fisheries Management Councils have been established, three of which manage the fish stocks on the Atlantic Seaboard. While the Fisheries Management Councils have been established primarily to manage stocks sought after by the commercial sector, some stocks, notably tuna, bluefish, sharks and cod are pursued by both the commercial and recreational fishery. As a result management regimes for certain stocks has been established including tuna.

Furthermore, in areas designated harbors by the Federal government, the US Army Corps of Engineers (USACE) may cede certain powers to the coastal state which generally transfers them to the local municipalities. These powers are generally limited to designating anchorage and mooring areas and the setting of speed limits within the designated harbors, all of which are intended to increase navigational safety.

From a marine recreational point of view, very few restrictions have been placed on the marine environment which, for purposes of transit, is almost exclusively governed by the Federal government through the commerce clause which protects free and innocent ocean passage even within the Territorial Sea.

The problems associated with land-based coastal facilities are considerably more confusing insofar as both the private and the public sectors have invested in these facilities. Historically, the public sector has only stepped in in those instances where a) the services provided were intended to serve a large proportion of the population, and b) where the private sectors were unable to fill the need because of prohibitive investments, or where special permits and/or licenses were required. Nonetheless, there are many instances where public marinas, boat ramps, fishing piers, and even beach facilities compete with private structures which provide essentially the same services. Some of these projects date back to the second Roosevelt administration's effort to get the country moving again following the economic depression which started in 1929 and which continued well into the 1930s. During this period several public works programs were initiated, which primary purpose was to employ people to build, revive, or clean abandoned and deteriorating physical environments.

3. Conclusions

The picture of marine recreational activities in North America, in general, and along the Eastern Seaboard in particular, is a confusing one because of a rapid growth in the number and financial wherewithall of many consumers. Another factor relates to the virtual explosion of new materials, new products, and new sports activities, most of which originate with the young.

While these trends are difficult to trace except in the rough, there is one additional overriding reason why coastal and marine recreational managers must be alert. New techniques and activities are constantly being tried out, and if successful, invariably will be adopted and become part of the marine recreational repertoire of the American public.

Marine recreational planners must incorporate in their plans facilities (environments) which will be able to serve the needs of such new marine recreational pursuits. The marine recreational system is not likely to remain static for very long. Such demands require the services of coastal managers with very unique talents. Without such efforts, the scarce coastal regime is not likely to be able to serve the needs of future marine recreational users.

Notes

1. This estimate includes persons who have passed one of several certification courses. However, this estimate is inflated in that many appear not to continue with the sport following actual certification.
2. While these developments are encouraging, it does not mean that nearshore water pollution in the U.S. has been solved. Major problems related to organic compounds and heavy metals will adversely affect the use of many estuaries and nearshore areas adjacent to urban areas.
3. In the context of this paper, accessibility refers to the ability and inability to travel from one's residence to the preferred beach and is primarily a function of limited transportation opportunities. Access has two meanings. Vertical access means the physical path from an existing public right-of-way (ROW), such as road or square to the beach. Horizontal access refers to the ability to travel parallel to the shoreline or water's edge.
4. An example of such efforts is the attempt to negotiate conflicts between the towboat operators and crab fishermen within Oregon's territorial sea. For further information, contact Oregon State University, Marine Advisory Service, Marine Science Center, Newport, OR.
5. In December 1988, shortly before leaving office, President Reagan extended the Territorial Sea from 3 to 12 miles by Presidential Proclamation. At the time of writing, the State jurisdiction remains at 3 miles.

References

Bascom, W. (1976) *Deep Water Ancient Ships*, Doubleday, New York, p. 72.

Bryan, H. (1977) The Sociology of Fishing: A Review and Technique, in Henry Clepper (ed.) Marine Recreational Fisheries, Proceedings of the First Annual Marine Recreational Fisheries Symposium, New Orleans, Louisiana, February 27, 1976 Sports Fishing Institute, Washington, DC.

Buchanan, (1972) as cited by Ditton and his co-workers in footnote 32, *Proceedings*, Sports Fishing Seminar, Tekyll Island, GA, p. 27. as cited by Ditten, Robert B. *et al.*, *Coastal Resources Management*, Lexington Books, Lexington, MA.

Cabelli, V. J., Morris, A. L. and Dufour, A. P. (1983) 'Public Health Consequences of Coastal and Estuarine Pollution: Infectious Diseases,' in E. P. Myers and Elizabeth Harding, *Ocean Disposal of Municipal Wastewater: Impacts on the Coastal Environment*, Vol. 2, MIT Sea Grant, Pub. MITSG 83–33, Cambridge, MA.

Cable, T. T., Udd, E. and Fridgen (1987), 'Protecting Anglers from Toxic Chemicals: An Interpretive Challenge,' paper accepted for the National Interpreters Workshop Meeting, Nov. 1–7.

Chaney, T. (1979) 'Coping with Boating Congestion in Maryland', unpub. paper, Maryland Department of Natural Resources, Coastal Zone Unit, Anapolis, MD.

Clawson, M. (1963) Land and Water for Recreation, Resources for the Future, Inc. and Rand McNalley and Comp., New York, NY.

David, E. (1971) 'Public Perceptions of Water Quality', *Water Reso Research* 7, 3, 453–457.

Dinius, S. H. (1981) Public Perceptions in Water Quality Evaluations,' *Water Resources Bull.* **17**, 1, 116–121.

Duell, D. (1977) Marine Recreational Fisheries Uses and Values in Coastal Recreational Resources in an Urbanizing Environment, University of Massachusetts and MIT, Boston, MA.

Foresta, R. A. (1984) *America's National Parks and Their Keepers*, Resources for the Future, Inc., Washington, DC.

Freeman, B. (1977) 'A Description of Recreational Finfish Along the Atlantic Coast in Relation to the Utilization of Living Marine Resources,' in Coastal Recreational Resources in an Urbanizing Environment, *op. cit.*

Gardozo, Y. and Hirsh, B. (1985) 'Florida Artificial Reefs – Alive and Well', *Sea Frontiers* **31**, 325–333.

Giesecke, A. G. (1987) 'Shipwrecks: The Past in the Present,' *Coastal Management* **15**, 3, 179–195.

Gordon, E. (1983) The Effects of Potable Water and Wastewater on the Development of two Coastal Communities,' unpub. MA thesis, Department of Marine Affairs, University of Rhode Island, Kingston, RI.

Hawthorne, N. (1879) *House of Seven Gables*, Houghton Mifflin, New York.

Heatwole, C. A. and West, N. (1983) 'Urban Shorebased Fishing, a Health Hazard?' *Proceedings*, CZ83, San Diego, CA, June 1–4, 2587–2598.

Heatwole, C. A. and West, N. (1984) 'Shorebased Fishing in New York City, Final Report, New York Sea Grant Program, New York, NY.

Heecock, R. D. (1966) Public Beach Recreation Opportunities and Patterns of Consumption on Cape Cod, unpub. Ph. D. Dissertation, Graduate School of Geography, Clark University, Worcester, MA.

Matheusik, M. R. E. and Mills, A. S. (1983) 'Sports Divers and Their Preferred Coastal Setting', CZ83, *Proceedings*, 1488–1506.

Mathieson, A. and Wall, G. (1982) Tourism: Economic, Physical and Social Impacts, Longman Publishers, London, Eng.

Matley, I. M. (1976) 'The Geography of International Tourism', Association of Amer. Geog., Resource Paper 76–1, Washington, DC.

Moser, G. (1987), 'Water Quality Perception, A Dynamic Evaluation', *Jour. of Environmental Psychology* **33**, 4, 259–262.

NMFS (1986) Marine Recreational Fishery Statistics Survey, Atlantic and Gulf Coast, 1985, Current Fishery Statistics 8327, U.S. Department of Commerce, Washington, DC.

Powell, S. C. (1963) *The Puritan Village*, Wesleyan University Press, Middletown, CT.

Robert, R. *et al.* (1986), 'Managing the Risks of Toxic Exposure, *Environment* **28**, 9, 19–37.

Scott, M. (ed.) (1985) The San Francisco Bay Area: A Metropolis in Perspective (2nd ed.), University of California Press, Berkeley, CA.

Sheehy, D. T. (1986) 'The Application of Designed Artificial Reefs in Coastal Mitigational Compensation and Fisheries Development Projects. The Coastal Society', *Briefing* **9**, 1, 14–17.

Spaulding, I. (1976) 'Sociocultural Values of Marine Recreational Fishing, in Marine Recreational Fisheries, *op. cit.*

West, N. (1978), 'Coastal Demographic Changes in the United States 1950–1976', *Proceedings*, Oceans 78, Washington, DC.

West, N. (1982) 'Marine Recreational Conflicts in the Nearshore Marine Environment', Proceedings, Marine Resource Management and The Future Role of the State of California, Asilomar, Pacific Grove, CA, November 7–10.

White, L. (1967) 'The Historical Roots of Our Ecological Crisis', *Science* **155**, 1203–1207.

Niels West
Department of Marine Affairs,
University of Rhode Island, U.S.A.

23. Beach resort morphology in England and Australia: A review and extension

Resort towns provide a special example of urban settlement that satisfies what Cohen calls the phenomenological 'recreational' mode of tourism, in which the holidaymaker escapes his usual environment and returns refreshed to his everyday life-world. Their landscapes are shaped to enhance that escape, and to facilitate it, leading to specialised urban morphologies which require separate examination. Resorts are commonly located close to Nature (Lavery, 1971) and Nature indeed is a key to understanding this morphology given its value as a social category in recent Western culture (Jeans, 1983; Tuzin, 1977). It is argued here that a common semiotic morphology can be imposed on seaside resorts, while recognizing international idiolects which do not disturb the general model.

Seaside recreation in England

Seaside resorts replaced the spa town as the fashionable place of recreation in the early nineteenth century, adopting the spa town's fashionable promenade in the process. Medical men were extolling the virtues of seawater, the upper classes wished to escape the middle class invasion of the spa towns and followed royalty, beginning with the Prince Regent at Brighton, to the seaside. The rise of railway travel made the seaside more widely accessible to a growing middle class, differentiated resorts came into existence – and the extension of cheap excursion rail travel and holidays with pay eventually opened the resorts to a wide section of the working class. Being beside the sea was, until after 1920, the direct aim in recreation, and even quite recently a survey showed that only one-quarter of visitors to one resort actually entered the water (Ashworth, 1984). The urban morphology of the English resort therefore has traditionally provided the chief resource used by holiday-makers (Pimlott, 1947; Cosgrove and Jackson, 1972).

Seaside recreation in New South Wales

Unlike England, resorts in Australia have, until the 1960s, been virtual suburbs of the large capital cities in which sixty per cent of the population has lived. Seabathing was popular in early New South Wales but evangelical disgust with the display of naked bodies led to it being banned in daylight hours in 1835. Resorts did develop, but following the English pattern of on-shore recreation until the law was successfully challenged in 1902–3. Regulations in other states were less strict, and bathing machines were introduced to Victoria and South Australia in 1885. In the

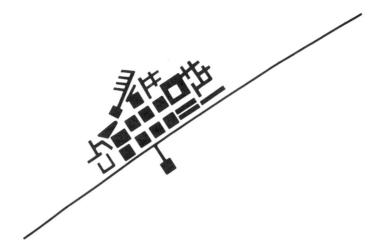

Figure 1. Morphology of a beach resort (after Gilbert, 1939).

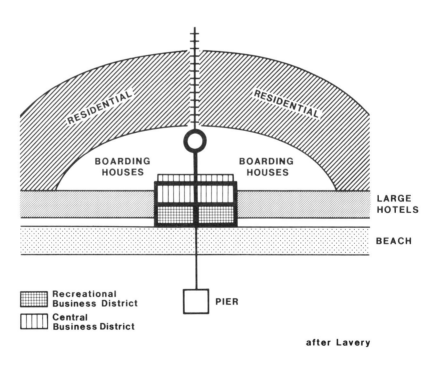

Figure 2. Morphology of a beach resort (Adapted from Lavery, 1971).

twentieth century, resort use has centred on going into the water, in bathing dress, and at all ages. Body-surfing and surf-board riding have become widespread skills. The beach has become a major feature of the Australian way of life, enjoyed throughout the summer by the majority of the urban population, rather than during planned episodes as in England (Wells, 1982).

The development of a geographical morphology of beach resorts

In 1939 E. W. Gilbert reviewed the history of resort development in England and made a primitive contribution to their morphological investigation (Figure 1). Gilbert published a number of such maps which showed the centrality of the pier and the gross plan of the built up area. He also published photographs of typical resort landscapes, but did not relate them to any spatial scheme. He did however point to the need for a flat and sandy shore, ignoring the pebbles of Brighton and the mud of Western-super-Mare (Gilbert, 1939).

A more sophisticated model was published by Lavery (Lavery, 1971) which is redrawn in Figure 2. Lavery's own drawing shows the beach as an insignificant, almost residual feature in a mainly urban landscape, though its characteristics are described as important to resort development. Behind the beach is the promenade, fronted by a row of large and expensive seaview hotels, behind which is a zone of cheaper boarding houses close to the centre of the resort. The outer zone is occupied by the ordinary residents of the resorts who perform service functions. Following Stansfield, Lavery identifies a recreational business district which serves holiday-makers with restaurants, and souvenir and other shopping facilities, located close to the water front and backed by the town's own CBD, since it usually also plays some larger role in a central place system. The railway station is shown as the chief means of access, which may have been outdated even in the England of 1971.

These analyses are concerned mainly with the landward side of the beach, which accords well with British holiday-makers' behaviour, but tends to neglect the beach and particularly the sea, while, being entirely morphological, they ignore the rich patterns of meanings which are built into the landscape. The aim now is to use semiotics to inject this pattern of meanings and so explain the morphology in terms of experience.

The semiotic morphology of the seaside resort

Figure 3 shows a morphology of the seaside resort that draws many features from Lavery. Inserted, however, is a zone of greenery, cliffs or dunes which commonly separate the promenade from the town, though not necessarily continuously. Road access, and the spread of boarding houses along such routes has been allowed for, and as commonly noticed higher class residential areas tend to occupy positions closer to the sea. The chief additions are however, two axes of meanings which structure morphology and behaviour within the resort.

The first axis runs perpendicularly to the land/sea junction. The basic distinction

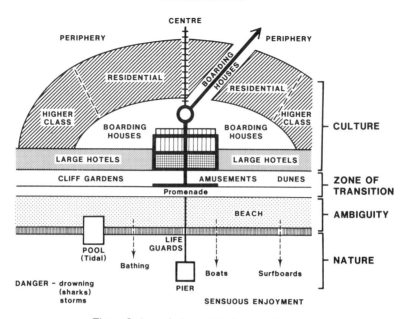

Figure 3. A semiotic model of the beach resort.

here is between the resort, representing Culture, and the sea, representing Nature. Culture is controlled and structured by class and function, and represents control, though elements of fantasy may be built into it, as with the Prince Regent's pavilion at Brighton, to signify its difference from the culture of the everyday working city. Resort advertising posters often play down this element, emphasizing rather the sea, the beach and the countryside surrounding the towns which is seen by Europeans as 'natural'. In Australian urban resorts, the CBD function is located in the metropolitan CBD, though Pigram (1977) has found that Lavery's morphology fits those east coast non-metropolitan Australian resorts he investigated.

The sea represents Nature to which man has been drawn since the eighteenth century under a multiplicity of influences such as the writings of J. J. Rousseau and the paintings of Phillipe de Loutherborg. The sea provides the tantalising dangers of nature – storms, drowning, and in Australia sharks, but also the compensating sensuous pleasures of relaxation, abandonment, enjoyment and the exercise of skills of survival. The attraction of this element has increased: early postcards (ca 1900) show few entering the water in English resorts, while now many do. In Australia entering the surf is a common experience, and one which has been given a special cultural meaning. Writing in the journal *Lone Hand* in 1910, a columnist said

> The surf is a glorious democracy – or rather it represents a readjustment of all the classifications that history and politics and social conditions ever brought about. Wealth has no place here, nor rank, nor scholarship. Plain primitive manhood and womanhood are the only tests the surf battler applies to distinguish one from another (Wells, 1982).

To the Australian Legend of equality the sea offers a special opportunity of realisation. Only recently are the private beaches found in parts of Europe coming to exist in Australia.

Between Culture and Nature however stands a zone of ambiguity. This may be provided by nature in the form of cliffs or unstable dunes, which may however be 'naturally' enhanced by plantings, or take a variety of forms from extensive grades to a simple strip of grass. This can be seen as a zone of transition from Culture to Nature, a symbolic transition zone transversed by signposted paths. Much more significant is the beach, which provides a zone of ambiguity between Culture and Nature. Here people behave and undress 'naturally', while culture attempts to maintain control.

Methods of control are emphasized by Fiske (1983), in the form of restrictions on littering, banning of dogs, access by vehicles, and clothing. In New South Wales, control was for seventy years exercised by banning sea-bathing, and since relaxation of the ban, there has been a continued battle to exert cultural control over the beach. At first only neck-to-knee costumes could be worn, and men's and women's bathing areas were segregated. Men's adoption of V-trunks had been forced upon the authorities by 1912. Then there was a scandal with the first appearance of the bikini in the 1950s – it was banned – and most recently topless women and even nude bathing have appeared on the beaches in defiance of authority. The current complaint is of open copulation taking place. Consistently authority has attempted to impose social control over the beach: consistently the beach-frequenter has defied and defeated such controls in an assertion of the rights of Nature in that domain. In Peircean triadic semiotics, the beach takes on the quality of 'thirdness', which can only exist and be defined in relation to two other signs, in this case Nature and Culture.

A special sign is the water's edge. Crossing it was a cultural project which took many years for the masses to achieve. Even now, observation of individual behaviour on crossing that divide shows an awareness of significance, even if there are diverse behaviours exhibited. A special machinery for doing so has been evolved, from the early bathing machines, which in Australia were provided with restricting shark-proof nets, to the modern surf and sail-board. Tidal pools are common in Australia as a safe way of crossing the divide, while in England, pools may provide an alternative to the divide.

A second axis of meaning runs parallel to the coast, where a centre is surrounded by peripheries at each end of the beach. In English resorts the centre is marked by the pier, which allows the public to penetrate nature in safety, but at the same time to experience the thrills of an apparently flimsy structure which gives access to the experience of being in Nature. There may be a sexual connotation of penetration but at the same time analysis of photographs of people on piers shows these much more used, as is the promenade, for looking at other people than at the sea. In Australia, some early 'piers' were built, as at Brighton and St. Kilda in Victoria, and Glenelg in South Australia, but these are more like jetties and the early English 'piers' than the grand edifices which developed at Southend and Brighton among other places. Rather the centre of the Australian beach is occupied by two flags, which designate the safe area for bathing which is patrolled and kept under observation for sharks by life savers. Here the family and older recreationists assemble.

The peripheries are occupied by the young and the non-conformist. Here a dog may be brought without detection. Here semi nude women and nude bathers and copulation first appear, though the first has now spread into the Australian centre, representing another loss to cultural authority.

The difference between centre and periphery on the beach can be seen in the existence of two groups which contrast in activities and values (Pearson, 1979). At the centre are the life savers, who value exercise, fitness and health, are socially affiliated to clubs patronised by public figures, who are highly competitive reflecting the values of the larger society, also authoritarian and emphasizing mastery over nature. Banned from the centre by fiat are the surfboard riders, who are individualistic, hedonistic, have informal social affiliations mostly with each other, and emphasize the sense of harmony with nature that their skills provide.

The latter group rejects the values of the central culture in attitudes, dress, and behaviour, and so falls into Cohen's 'diversionary' group, who seek recreation as a form of escape from society felt to be oppressive. They see the lifesavers as conformist, arrogant, and conservative. Since Pearson wrote, surf-board riding has been colonised by the introduction of sponsored competitive events with large prize money and an international circuit. The riders are becoming more competitive and conventional, and this illustrates the larger role of the periphery in capitalist society. Unable to generate many new forms of consumption within itself, the centre colonises the periphery, which must continue to innovate non-central behaviour anew: in this way, the periphery can be seen as essentially functional within capitalist society.

The pier and the life-savers have their central counterparts in the morphology of the resorts, in a coinciding alignment of recreational business district and transport arrangements, reflecting the 'ice-cream vendor solution' in locational geography. The whole morphology depends upon a set of meanings translated into behaviour, so that such a model as Figure 3 should provide a basis for behavioural research in resort areas.

For example, we may ask how far from the beach people are willing to wander in bathing dress, and find spatial limitations to this area which have tended to expand over time and vary with age. Public undressing may be valid in some areas, while the use of sex-segregated changing rooms may be required in others. The behaviours of controllers such as Australian beach inspectors, and their values, may be investigated. Behaviour and morphology have grown together, and continually change, while the existence of international beach idiolects, slightly explored here, may be a rich field for investigation.

In Barthesian terms, the elements of the seaside resort constitute a systemic set of signs which are replicable in different resorts, while the juxtaposed signs which constitute the single resort may be seen in syntagmatic relationship (Barthes, 1967). The effect of the syntagm is 'neutralization', since the dangerous sea of Nature is rendered safe in the resort, while retaining its dangerous qualities on the wild coast. The syntagm represents a familiar possession, the inhuman state of nature is denied, alienation is refuted, nausea is circumvented. The resort is a structure of

Figure 4. The morphological zones of a beach resort, indicating the sea as nature, the beach as a zone of ambiguity, the vegetation-enhanced zone of cliff transition, the zone of large hotels, and the zone of boarding houses. Bournemouth in the 1930s.

Figure 5. The recreational mode of seaside recreation is sometimes reflected in fanciful architecture. Entrance to the pier at Bournemouth in the 1890s.

Figure 6. Even after walking upon a sunlit pier in smooth seas, visitors might wish to record their experience by purchasing a postcard indicating the pier defying the sublime forces of Nature. The pier at Hunstanton, 1905.

Figure 7. Beach at Southsea, early 1900s. Few enter the water, many however gather around its edge. A bathing machine is provided for serious participants in Nature. Zone of large hotels and the pier are visible.

information, introducing a poetics which cannot be described by plans, but can be seen in illustrations (see Figures 4, 5, 6, 7). As Barthes has written, 'the object is the world's human signature' (Barthes, 1980).

References

Ashworth, G. S. 1984, *Recreation and Tourism,* Bell and Hyman, London.
Barthes, R. 1980, *New Critical Essays* (transl). Hill and Wang, New York, p. 24.
Barthes, R. 1967, *Elements of Semiology,* Hill and Wang, New York,
Cohen, E. 1979, 'A phenomenology of tourist experience', *Sociology* 13, 179–201.
Cosgrove, I. and Jackson, R. 1972, *The Geography of Leisure and Recreation,* Hutchinson, London, Chapter 2.
Fiske, J. 1983, 'Surfalism and Sandiotics: the beach in OZ popular culture', *Austr. Jnl. of Cult. Stud.* 1, 120–149.
Gilbert, E. W. 1939, 'The growth of inland and seaside health resorts in England', *Scott. Geogr. Mag.* 55, 16–35.
Jeans, D. N. 1983, 'Wilderness, Nature and Society: contributions to the history of an environmental attitude', *Austr. Geogr. Stud.* 21, 170–182.
Lavery, P. (ed.) 1971, *Recreational Geography,* David and Charles, Newton Abbot.
Pimlott, J. A. R. 1947, *The Englishman's Holiday: a social history,* Faber, London.
Pigram, J. J. 1977, 'Beach resort morphology', *Habitat Intnl.* 2, 525–541.
Tuzin, D. F. 1977, 'Reflections of being in Arapesh water symbolism', *Ethnos* 195–223.
Wells, L. 1982, *Sunny Memories. Australians at the Seaside,* Greenhouse, Richmond Vic., p. 140.

Dennis N. Jeans
Department of Geography,
University of Sydney,
Australia

The GeoJournal Library

1. B. Currey and G. Hugo (eds.): *Famine as Geographical Phenomenon.* 1984
ISBN 90-277-1762-1
2. S. H. U. Bowie, F.R.S. and I. Thornton (eds.): *Environmental Geochemistry and Health.* Report of the Royal Society's British National Committee for Problems of the Environment. 1985　　ISBN 90-277-1879-2
3. L. A. Kosiński and K. M. Elahi (eds.): *Population Redistribution and Development in South Asia.* 1985　　ISBN 90-277-1938-1
4. Y. Gradus (ed.): *Desert Development.* Man and Technology in Sparselands. 1985　　ISBN 90-277-2043-6
5. F. J. Calzonetti and B. D. Solomon (eds.): *Geographical Dimensions of Energy.* 1985　　ISBN 90-277-2061-4
6. J. Lundqvist, U. Lohm and M. Falkenmark (eds.): *Strategies for River Basin Management.* Environmental Integration of Land and Water in River Basin. 1985　　ISBN 90-277-2111-4
7. A. Rogers and F. J. Willekens (eds.): *Migration and Settlement.* A Multiregional Comparative Study. 1986　　ISBN 90-277-2119-X
8. R. Laulajainen: *Spatial Strategies in Retailing.* 1987　　ISBN 90-277-2595-0
9. T. H. Lee, H. R. Linden, D. A. Dreyfus and T. Vasko (eds.): *The Methane Age.* 1988　　ISBN 90-277-2745-7
10. H. J. Walker (ed.): *Artificial Structures and Shorelines.* 1988
ISBN 90-277-2746-5
11. A. Kellerman: *Time, Space, and Society.* Geographical Societal Perspectives. 1989　　ISBN 0-7923-0123-4
12. P. Fabbri (ed.): *Recreational Uses of Coastal Areas.* A Research Project of the Commission on the Coastal Environment, International Geographical Union. 1990　　ISBN 0-7923-0279-6
13. L. M. Brush, M. G. Wolman and Huang Bing-Wei (eds.): *Taming the Yellow River: Silt and Floods.* Proceedings of a Bilateral Seminar on Problems in the Lower Reaches of the Yellow River, China. 1989　　ISBN 0-7923-0416-0
14. J. Stillwell and H. J. Scholten (eds.): *Contemporary Research in Population Geography.* A Comparison of the United Kingdom and the Netherlands. 1990
ISBN 0-7923-0431-4
15. M. S. Kenzer (ed.): *Applied Geography.* Issues, Questions, and Concerns. 1989　　ISBN 0-7923-0438-1
16. D. Nir: *Region as a Socio-environmental System.* An Introduction to a Systemic Regional Geography. 1990　　ISBN 0-7923-0516-7